Immunology

Foundations and Frameworks Series

A Clear Guide to the Immune System, Disease, and Defense

Marcus Elwood

TME Press email: support@tme.net

Introduction

Break the skin with a paper cut and something starts happening before you even notice the sting. Bacteria that live on everyday surfaces slip into the opening. Within seconds, cells already stationed in your skin detect them and release chemical signals. Blood vessels widen. More immune cells travel toward the spot. Redness, warmth, and a little swelling show up over the next few hours. Most of the time you never think about any of it, and the cut heals on its own. That quiet, automatic response is your immune system doing its job, and it repeats thousands of times across your life without ever asking for your attention.

Your immune system runs day and night. Your bone marrow produces hundreds of billions of new blood cells every day, and a large share of them are white blood cells built for defense. One type alone, the neutrophil, makes up more than half of the white blood cells in your bloodstream, and it circulates for only a day or so before being replaced. The system stays in constant motion because the threats never stop. You breathe in microbes, swallow them with your food, and pick them up from every surface you touch. Most of them cause no harm at all, because your body handles them long before you feel anything.

The immune system also solves a hard problem. It has to recognize an almost unlimited variety of threats, including bacteria and viruses that have never existed before, while leaving your own cells completely alone. To do this, your body can produce more than a billion different antibody shapes, each one able to match a different target. Get this recognition wrong in one direction and the system attacks your own tissue. Get it wrong in the other direction and dangerous microbes slip through. Health, for the most part, depends on getting it right, over and over, in the background.

Understanding how this works changes how you see your own health. When you catch a cold, run a fever, get a vaccine, or hear that a friend has been diagnosed with an autoimmune condition, the immune system sits at the center of the story. News about cancer treatments, allergy shots, organ transplants, and new vaccines all trace back to the same set of principles. Once those principles make sense to you, the headlines stop sounding like jargon and start telling you something useful.

This book is written for anyone who wants a clear starting point. You don't need a background in biology or medicine. Students taking a first course, healthcare workers who want a solid refresher, and curious readers who simply want to understand their own bodies will all find a path through the material. Each idea builds on the one before it, and every technical term is explained in plain language the first time it appears.

The book moves in a deliberate order, from the ground up.

The first two chapters build the foundation. Chapter 1 explains what the immune system actually does, how it tells your own cells apart from foreign material, and where its organs and tissues sit inside the body. You will see how the whole system traces back to stem cells in your bone marrow. Chapter 2 introduces the cells themselves, the individual cells that carry out defense. You will meet neutrophils, macrophages, dendritic cells, T cells, B cells, and natural killer cells, and you will learn what each one does and when it shows up.

Chapters 3 and 4 cover the two main parts of the response. The innate response, in Chapter 3, is the fast, general reaction you are born with. It includes your skin and mucous membranes, the inflammation that follows an injury, and the complement system that tags microbes for destruction. The adaptive response, in Chapter 4, is slower but far more precise. It learns the exact shape of a threat, attacks that specific target, and stores a memory of it for next time. This memory is the reason you rarely catch the same illness, such as chickenpox, twice.

Chapter 5 goes deeper into antibodies and antigens, the molecules at the heart of precise defense. You will learn how an antibody is built, how it locks onto one target and ignores everything else, and why there are different classes of antibody for different jobs. This chapter also explains the difference between the antibodies your body makes on its own and the monoclonal antibodies now used to treat cancer, autoimmune disease, and other conditions.

Chapter 6 shows the immune system at work against real threats. Viruses, bacteria, fungi, and parasites each attack the body in their own way, and each one calls for a different response. You will learn why some viruses, like HIV and influenza, are so hard to fight, how bacteria set off your defenses, and why a fever is a planned strategy rather than a simple side effect of being sick.

Chapter 7 turns to what happens when the system breaks down. Sometimes it attacks the body's own tissue, which produces autoimmune conditions such as type 1 diabetes and rheumatoid arthritis. Sometimes it overreacts to harmless things like pollen or peanuts, which we experience as allergies. Sometimes it runs too weakly and leaves the body open to infection. This chapter also covers the link between long-term inflammation and diseases like heart disease, and the role the immune system plays in spotting and destroying early cancer cells before they grow.

Chapter 8 explains vaccination. Few medical tools have saved more lives. Smallpox alone killed an estimated 300 million people in the twentieth century, and a global vaccination effort wiped it out completely by 1980. You will learn how a vaccine trains your immune system without making you sick, how live and inactivated vaccines differ, and how the newer mRNA vaccines work. The chapter also covers herd immunity, the testing every vaccine goes through before it reaches you, and the worldwide programs that have pushed diseases to the edge of extinction.

Chapter 9 looks at where immunology is headed. You will read about cancer immunotherapy, which trains the body to attack tumors it once ignored, and the challenge of stopping the immune system from rejecting a transplanted organ. You will also learn how the trillions of microbes living in your gut help regulate your immune system, and how doctors are starting to tailor treatment to each person's individual immune profile.

Several sections after the final chapter support your reading. A historical timeline traces the discoveries that built the field, starting with the first smallpox vaccine and running through the gene-based therapies of today. Every chapter ends with review questions, and a dedicated section provides example answers so you can check your understanding as you go. A section of key terms and definitions gathers the vocabulary in one place for quick reference. A closing list of works cited and recommended readings points you toward deeper study whenever you want to keep going.

By the end, you'll understand the immune system as a working whole rather than a scattered list of parts. You will know why you get sick, why you get better, and why some illnesses never come back. You will understand what a vaccine is doing inside your body, why allergies happen, and how an immune system can turn against the person it is meant to protect. That understanding has real, practical value. It helps you follow medical advice with more confidence, ask sharper questions at the doctor's office, and read health news without feeling lost.

The immune system protects you every hour of every day, usually without a single sign that it's working. This book gives you a clear look at how it does that, one piece at a time.

Outline

Contents

Chapter 1: Foundations of the Immune System

Think about the last time you had a sore throat, a scraped knee, or a stomach bug that passed in a day. In each case, something inside you noticed a problem and went to work fixing it. You didn't decide to fight the infection. You didn't tell your body to heal the cut. All of that happened on its own, run by a system most people never see and rarely think about.

The immune system is that hidden manager. It watches over every part of your body at once, sorts helpful from harmful, and responds to trouble faster than you can feel it. It's not one organ you can point to, like the heart or the liver. It's a spread-out network of cells, tissues, and organs that communicate constantly and cover you from head to toe.

This chapter lays the groundwork for everything that follows. Before we look at individual cells or specific diseases, you need a clear picture of what the immune system is for, how it tells friend from foe, the two broad strategies it uses, where its machinery lives, and where all of its cells come from in the first place. Get these five ideas straight, and the rest of the book will click into place.

What the Immune System Actually Does

Most people describe the immune system in one sentence: it fights germs. That's true, but it undersells the job by a wide margin. Protection against bacteria and viruses is the part you notice, because that's when you feel sick. The system does a great deal more that you never register at all.

Start with the scale of the threat. Your body is a warm, wet, nutrient-packed environment, which makes it an ideal place for microbes to grow. Bacteria, viruses, fungi, and parasites would love to move in. Some of them can double their numbers every twenty minutes under good conditions, so a handful of cells becomes millions in a matter of hours. If nothing stopped them, a minor infection would overwhelm you in a day or two. The immune system is the reason that doesn't happen.

But defense is only the first of three main jobs. The second is cleanup. Your own cells are constantly wearing out, getting damaged, and dying on schedule. This programmed cell death, called apoptosis, happens on a scale that's hard to picture. Tens of billions of your cells die every single day as part of normal maintenance, and something has to clear away the debris before it piles up and causes trouble. Immune cells handle most of that removal quietly, hauling off dead cells and broken bits without any inflammation or symptoms. When you're bruised and the purple mark slowly fades to yellow and disappears, you're watching immune cells clean up damaged tissue and broken-down blood.

The third job is surveillance. Every day, some of your cells make mistakes when they copy their DNA to divide. A few of those mistakes turn a normal cell into a potential cancer cell, one that could grow without limit if left alone. Part of your immune system patrols for exactly these rogue cells and destroys them before they can form a tumor. This quiet policing goes on for your entire life, and most of us never know when it catches something.

So the honest description of the immune system is broader than germ-fighting. It defends, it cleans, and it patrols. All three run at the same time, all the time, across every tissue you have.

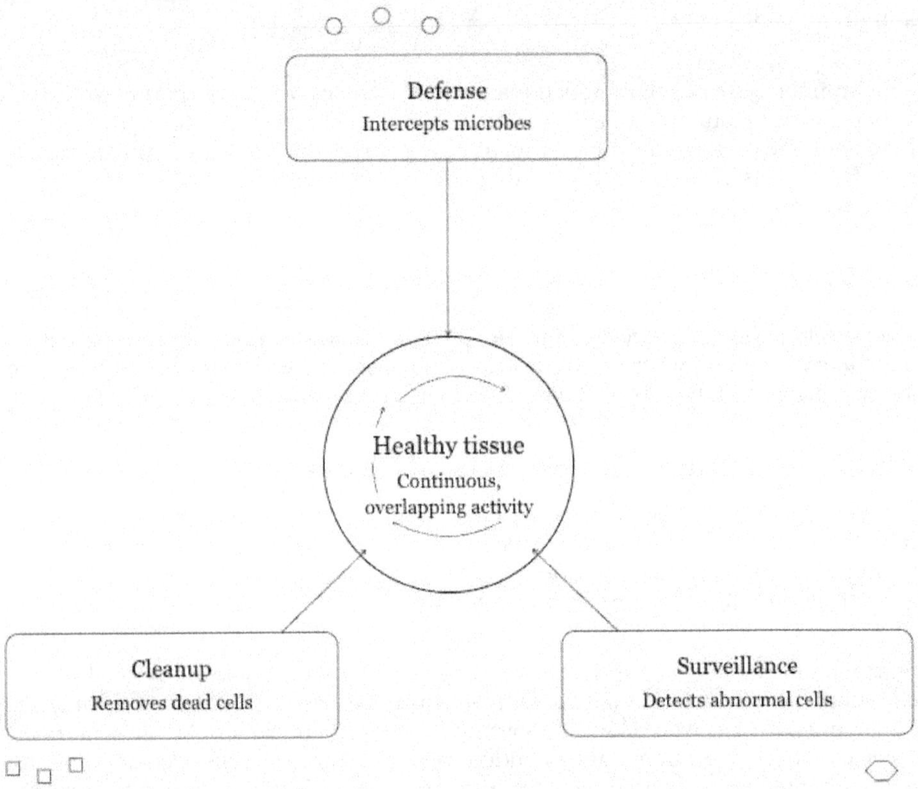

Here's a point that surprises a lot of newcomers. The immune system also has to know when to do nothing. Your skin, mouth, and gut are home to enormous populations of bacteria that help you rather than harm you. Your large intestine alone carries hundreds of species of microbes that break down food you couldn't digest on your own, make certain vitamins, and crowd out dangerous invaders. Your immune system lives right next to all of them and mostly leaves them in peace. Learning to tolerate the helpful residents is just as important as attacking the harmful ones. An immune system that overreacts is as dangerous as one that underreacts, which is a theme you'll see again and again in this book.

Consider the physical territory this system has to cover. Your skin spreads across roughly two square meters and forms your outer wall. That's the surface everyone thinks of. But your inner surfaces are far larger. The lining of your gut, if you could unfold all its ridges and folds, covers something like thirty square meters. Your lungs, with their millions of tiny air sacs, add even more. Every one of these surfaces meets the outside world, and every one of them is a place where microbes try to get in. The immune system has to guard all of it at once, which is why its cells are scattered throughout your body instead of concentrated in one place.

The response also has to match the threat. A tiny splinter in your finger calls for a small, local reaction. A serious bloodstream infection calls for a full-body emergency response with fever, fatigue, and a flood of immune cells. The system reads how big and how dangerous a problem is, then dials its reaction up or down to fit. Overreact to a splinter and you'd be miserable over nothing. Underreact to a real infection and you could die. Getting the size of the response right is one of the most impressive things the immune system does, and it manages it without any conscious input from you.

There's a cost to all this activity, and it's worth naming. Running an immune response burns energy and can damage your own tissue in the process. The swelling, heat, and pain of an infected wound come partly from the fight itself, not just the microbes. That's why the system doesn't stay at full alert all the time. It idles at a low level, watching and waiting, and ramps up only when it detects a real problem. Balancing readiness against cost is a constant trade-off, and the whole system is tuned to strike it.

Self vs. Non-Self Recognition

Everything the immune system does depends on one basic skill: telling the difference between what belongs in your body and what doesn't. Biologists call this distinguishing self from non-self. It sounds simple when you say it in plain words. In practice, it's the hardest problem the immune system solves, and getting it wrong in either direction causes serious disease.

Think about what the system is up against. It has to recognize and attack bacteria, viruses, fungi, and parasites, including brand-new ones it has never encountered, while leaving your trillions of healthy cells completely untouched. Your own cells and a bacterial cell are both made of proteins, fats, and sugars, built from the same basic chemistry. The immune system can't just look for "life" and attack it, because you're alive too. It needs a way to read a molecular ID card and decide, in a fraction of a second, whether the cell in front of it is yours or a stranger.

Your cells carry exactly that kind of ID. Nearly every cell in your body displays a set of surface proteins that act as identification tags. In humans, the main tags come from a group of genes called the human leukocyte antigen system, or HLA, which is the human version of what scientists more broadly call the major histocompatibility complex, or MHC. These proteins sit on the outside of your cells and constantly show samples of what's happening inside. Immune cells patrol past, read the tags, and confirm that everything looks like it belongs. When a cell displays the proper self markers and normal contents, the immune system passes it by. When something looks foreign, the system takes a closer look.

What makes HLA so good at this is its staggering variety. The HLA genes are the most variable genes in the entire human genome. There are thousands upon thousands of different versions spread across the human population, and the particular combination you carry is almost certainly different from anyone you'll ever meet, unless you have an identical twin. Your exact set of HLA tags is nearly a personal signature. This is why organ transplants are so difficult. When a surgeon puts a donor kidney into your body, your immune system reads the donor's HLA tags, sees that they don't match your own, and attacks the new organ as foreign. Doctors spend enormous effort matching donors and recipients as closely as possible and still have to suppress the immune system with drugs to keep it from rejecting the transplant. Identical twins are the exception. Because they share the same HLA genes, one twin can donate an organ to the other with far less risk of rejection.

You can see self versus non-self recognition at work in something as ordinary as blood types. Your ABO blood group, the system that sorts people into type A, B, AB, or O, comes down to sugar molecules on the surface of your red blood cells. If you're type A, your immune system treats the type B sugar as foreign, and vice versa. Give someone the wrong blood type in a transfusion and their immune system attacks the incoming red cells immediately, which can be fatal. An Austrian scientist named **Karl Landsteiner** (left) worked this out in 1901, and it earned him a Nobel Prize in 1930. Blood typing before a transfusion is really just a way of respecting each patient's self and non-self boundaries so the immune system doesn't turn on the donated blood.

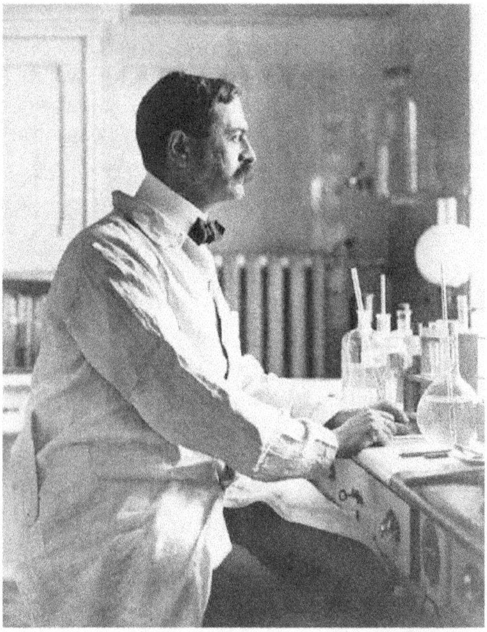

Now here's where the story gets more interesting. For most of the twentieth century, scientists thought of the immune system purely in terms of self versus non-self. Anything foreign got attacked; anything self got a pass. But that model has a hole in it. If the immune system attacked everything foreign, it would go to war against every harmless particle of dust you inhale, every bit of food you eat, and every helpful bacterium in your gut, all of which are technically non-self. Yet it doesn't. Something more subtle is going on.

In the 1990s, an immunologist named Polly Matzinger proposed a different way to think about it. She argued that the immune system doesn't respond to foreignness alone. It responds to danger. Damaged and stressed cells release chemical alarm signals, and it's those signals, not simply the presence of something foreign, that wake the immune system up. A harmless foreign protein that causes no damage may be ignored. A foreign invader that injures your tissue triggers a full response. This "danger model" doesn't completely replace the self versus non-self idea, but it fills in the gaps. The modern view combines both: the immune system watches for things that are foreign and for signs that your own cells are under attack, and it weighs both kinds of evidence before it acts.

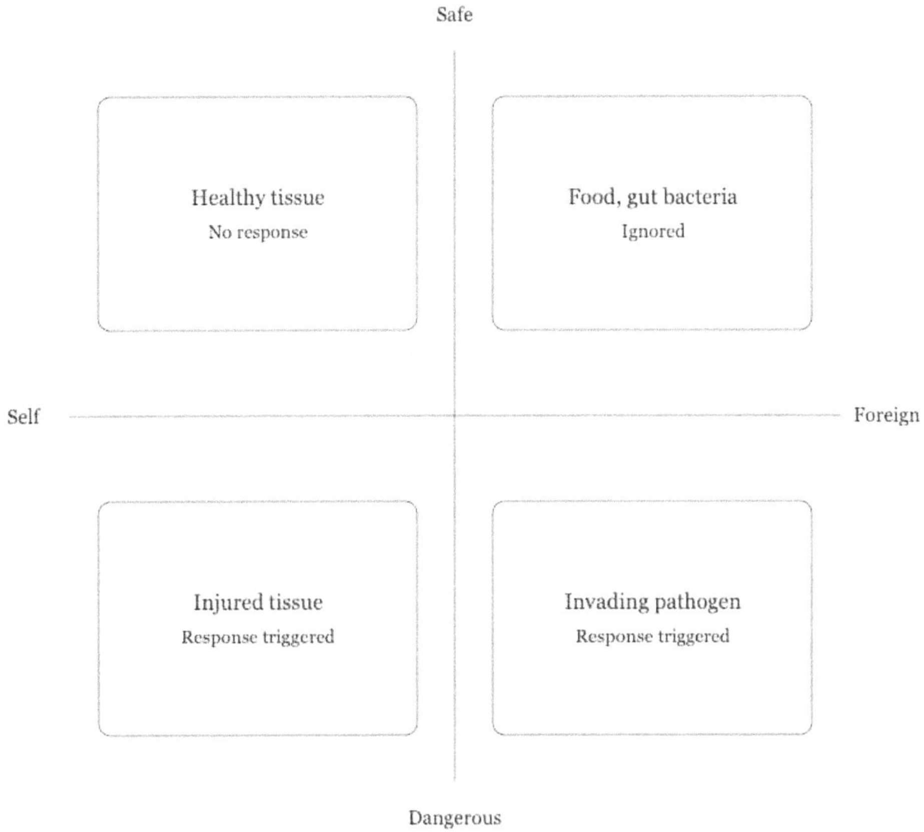

Activation requires identity plus evidence of danger

Getting this recognition wrong has a name in each direction. When the immune system fails to attack a real threat, infection or cancer can take hold. When it mistakenly attacks your own healthy tissue, treating self as non-self, the result is autoimmune disease. Conditions like type 1 diabetes, rheumatoid arthritis, and multiple sclerosis all come from an immune system that has lost the ability to recognize certain parts of the body as its own and attacks them by mistake. The stakes of self versus non-self recognition, in other words, aren't abstract. They're the difference between health, infection, and self-inflicted damage.

Innate vs. Adaptive Immunity

Your immune system runs on two connected strategies, and understanding the split between them is one of the most useful things you can learn in immunology. The two branches are called innate immunity and adaptive immunity. They work differently, run on different timelines, and handle different parts of the job. Once you can tell them apart, a huge amount of immune behavior starts to make sense.

Innate immunity is the part you're born with, ready to go from day one. It's fast, general-purpose, and doesn't need any prior experience with a threat to respond. Within minutes to hours of an invader breaking in, the innate system is already fighting. It includes your physical barriers like skin, the immune cells that gobble up bacteria on contact, and the chemical signals that trigger inflammation. The catch is that innate immunity isn't specific. It recognizes broad categories of danger, the general molecular patterns shared by whole classes of microbes, rather than any single invader in detail. It responds the same way to the same category of threat every time, whether it's your first encounter or your hundredth. And it keeps no memory of what it has seen.

Adaptive immunity is the opposite in almost every way. It's slower to start, extremely specific, and it learns. When a new threat appears, the adaptive system takes time to study it, custom-build a response, and deploy weapons aimed at that exact invader and nothing else. The first time your body meets a particular microbe, the adaptive response takes several days to gear up, usually somewhere between four and seven days for a first-time reaction. That delay is why a new infection can make you feel sick for a while before you recover. But the adaptive system makes up for its slow start with two things innate immunity lacks: precision and memory. It targets the specific invader with great accuracy, and once the fight is over, it keeps a record. The next time that same microbe shows up, the adaptive system recognizes it instantly and crushes it before you ever feel sick. This memory is the reason you usually catch certain illnesses only once, and it's the entire basis of how vaccines work.

The two systems are in constant communication, and the handoff between them is one of the most important events in an immune response. The innate system responds first, contains the threat as best it can, and then alerts the adaptive system, passing along samples of the invader so the slower, more precise branch can build its custom response. Certain cells act as messengers between the two, carrying information from the front lines back to the specialized adaptive cells. Neither branch is complete on its own. Innate immunity buys time and raises the alarm; adaptive immunity delivers the targeted, lasting response. You need both.

A comparison side by side makes the differences easier to hold in your head.

Feature	Innate Immunity	Adaptive Immunity
Present from birth	Yes, ready immediately	Develops over time with exposure
Speed of response	Minutes to hours	Days on first exposure
Specificity	General, recognizes broad categories	Highly specific to one target
Memory	None; responds the same each time	Yes; remembers past threats

Main players	Barriers, phagocytes, inflammation, complement	T cells, B cells, antibodies
Response on repeat exposure	Same as the first time	Faster and stronger each time

The evolutionary backstory adds a layer worth knowing. Innate immunity is ancient. Nearly every living thing has some version of it. Insects, worms, and other invertebrates rely on innate defenses alone and do perfectly well, surviving in filthy environments teeming with microbes without any adaptive immune system at all. Even plants have innate-style defenses that recognize and respond to invading microbes. This kind of immunity has been around for hundreds of millions of years because it works.

Adaptive immunity is the newer invention. It appeared in jawed vertebrates roughly 500 million years ago, which means the fish, amphibians, reptiles, birds, and mammals that descended from them all inherited it. Adaptive immunity was a major upgrade, adding memory and precision on top of the innate defenses that came before. But it never replaced the older system. Instead, evolution stacked the new capability on top of the old one, which is why you carry both. Your innate immunity is the deep, shared inheritance you have in common with a housefly. Your adaptive immunity is the more recent, more sophisticated layer you share only with other vertebrates.

Understanding which branch is doing what helps you make sense of your own experiences. The redness and swelling around a fresh cut is innate immunity acting fast and locally. The lasting protection you get from a childhood vaccine is adaptive immunity remembering. When you shake off a cold in a week, you're watching the adaptive system take its several days to identify the virus and then finish it off. Almost every immune event you'll ever notice is one branch, the other, or the two of them working together.

Major Organs and Tissues of Immunity

Since the immune system isn't a single organ, it helps to know where its parts actually live. Immune cells travel throughout your body, but they're made, trained, and stationed in specific places. Immunologists sort these locations into two groups: primary lymphoid organs, where immune cells are born and mature, and secondary lymphoid organs, where those cells gather, meet invaders, and launch responses.

The two primary lymphoid organs are your bone marrow and your thymus. Bone marrow, the soft tissue inside your bones, is the factory where nearly all of your immune cells are produced. It's also where one important type, the B cell, finishes maturing. We'll come back to the marrow in detail in a moment, since it's where the whole story begins. The thymus is a small organ sitting behind your breastbone, just above your heart. It's the training school for T cells, which are named for it. Young T cells travel to the thymus to learn a strict lesson: attack foreign threats, but never attack your own body. Cells that fail this test, the ones that would turn against your own tissue, get eliminated before they're ever released. Only the cells that pass graduate into circulation.

The thymus has an unusual life story. It's most active during childhood, when your body is building its T cell repertoire from scratch. After puberty, it gradually shrinks and is slowly replaced by fatty tissue, a process called involution. By the time you're elderly, your thymus is a small remnant of what it once was. This is one reason the immune system tends to weaken with age. The school that trains new T cells winds down, and your body leans more heavily on the T cells it produced earlier in life.

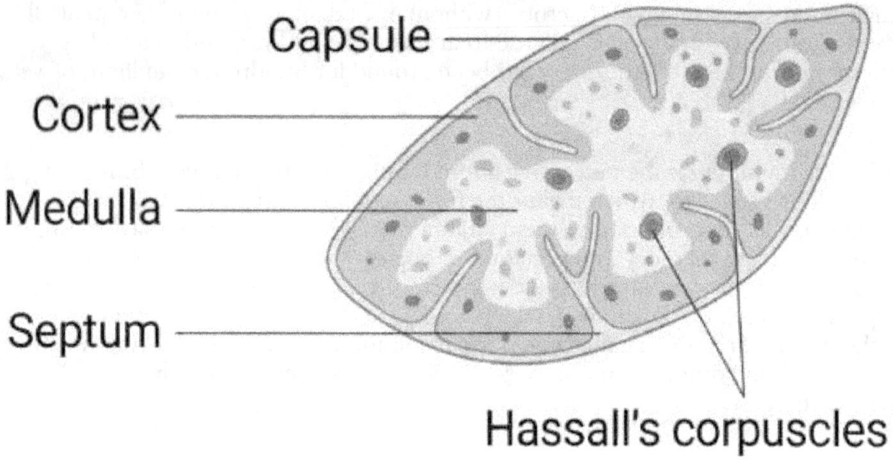

(The histology of the thymus. The **capsule** is the outer connective-tissue covering that protects the thymus. **Septa** extend inward, dividing it into lobules and carrying blood vessels. The **cortex** contains densely packed immature T cells, where early development and positive selection occur. The **medulla** contains more mature T cells and helps eliminate self-reactive cells. **Hassall's corpuscles**, formed by thymic epithelial cells, support T-cell maturation and immune tolerance.)

The secondary lymphoid organs are where the action happens once cells are mature. Your spleen, tucked under your ribcage on the left side and roughly the size of your fist, filters your blood. It catches microbes and worn-out blood cells passing through and gives immune cells a place to encounter bloodborne threats. Your lymph nodes, small bean-shaped structures numbering in the hundreds throughout your body, filter a fluid called lymph. When you've been sick and felt tender lumps in your neck, armpits, or groin, those were lymph nodes swollen with immune activity, working hard against a nearby infection.

Lymph nodes connect to a whole drainage network called the lymphatic system. Lymph is the clear fluid that seeps out of your bloodstream into your tissues, bathes your cells, and then gets collected and returned to your blood through a system of vessels. Along the way, it passes through lymph nodes, which act as checkpoints. Immune cells stationed in the nodes sample the passing lymph for any sign of infection. If they find something, they trigger a response right there. This is why the lymphatic system matters so much: it's the surveillance and transport network that lets immune cells patrol tissue far from any blood vessel.

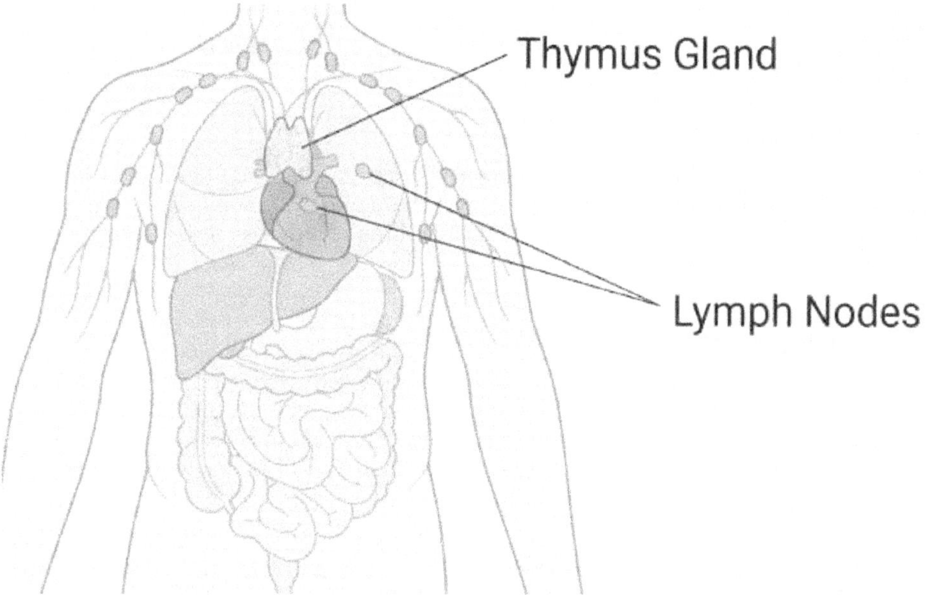

You also carry immune tissue in the places most exposed to the outside world. Your tonsils, at the back of your throat, catch microbes you breathe in or swallow. Patches of immune tissue in your intestinal wall, called Peyer's patches, guard the gut, where the microbial population is densest. Collectively, the immune tissue lining your gut, lungs, and other surfaces is known as mucosa-associated lymphoid tissue. It puts immune defenses exactly where invaders are most likely to try to enter.

Here's a summary of the main organs and what each one does.

Organ or Tissue	Type	Main Function
Bone marrow	Primary	Produces nearly all immune cells; matures B cells
Thymus	Primary	Trains and matures T cells; screens out self-reactive ones
Spleen	Secondary	Filters blood; responds to bloodborne threats
Lymph nodes	Secondary	Filter lymph; hubs for launching immune responses

Tonsils	Secondary	Catch inhaled and swallowed microbes at the throat
Peyer's patches	Secondary	Guard the intestinal lining against gut invaders

Seen together, these organs form a connected system with a clear division of labor. Some build and train the cells. Others deploy them and give them places to meet the enemy. And a transport network ties the whole thing together so that no corner of your body goes unwatched.

The Bone Marrow and Cell Origin

Every immune cell in your body traces back to the same starting point: the bone marrow. Deep inside your bones, in that spongy red tissue, sits a small population of remarkable cells called hematopoietic stem cells. These are the ancestors of your entire blood and immune system, and everything the rest of this book describes ultimately begins with them.

Hematopoietic stem cells are rare. They make up only a tiny fraction of the cells in your marrow, far less than one in a thousand. But they have two abilities that make them irreplaceable. First, they can renew themselves, dividing to make more copies of themselves so the supply never runs out over your lifetime. Second, they can turn into any type of blood cell your body needs. This process of generating blood cells is called hematopoiesis, and it runs continuously from before birth until the day you die.

When a hematopoietic stem cell commits to becoming a mature cell, it heads down one of two main paths. Immunologists call these the myeloid lineage and the lymphoid lineage. Think of them as two family branches, each producing a different set of immune cells.

The myeloid lineage produces most of the innate immune cells, the fast responders. This branch gives rise to the cells that swallow bacteria, drive inflammation, and handle the first wave of defense. It also produces your red blood cells and the platelets that clot your blood, so the myeloid branch does double duty, supplying both immune and non-immune cells.

The lymphoid lineage produces the cells of adaptive immunity, along with one important innate cell. This branch gives rise to T cells and B cells, the specialists that provide targeted, memory-based defense, as well as natural killer cells, which patrol for infected and cancerous cells. When you hear the word lymphocyte, it refers to a cell from this lineage.

Here's how the two branches break down.

Lineage	Produces	Role
Myeloid	Neutrophils, macrophages, dendritic cells, and others	Mostly fast innate defense; also red cells and platelets
Lymphoid	T cells, B cells, natural killer cells	Targeted adaptive defense and cell surveillance

The scale of this production line is genuinely hard to grasp. Your marrow runs around the clock, replacing cells as fast as they're used up. Many immune cells live only days before they die and get replaced, so the factory can never shut down. If your bone marrow stopped working, you'd run out of functioning immune cells within weeks and become dangerously vulnerable to infection. This is exactly what happens to patients whose marrow is damaged by disease or by treatments like chemotherapy, which is why doctors watch blood cell counts so closely during cancer treatment.

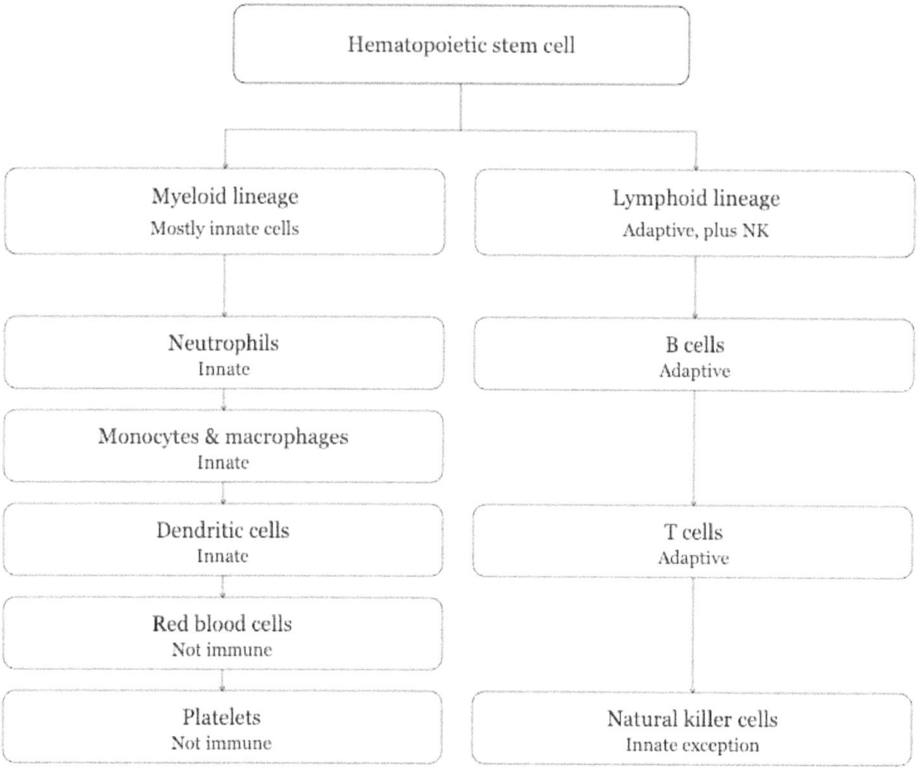

The central place of bone marrow also explains why bone marrow transplants can be lifesaving. In certain blood cancers, like leukemia, the marrow starts producing faulty cells. Doctors can wipe out the diseased marrow and replace it with healthy hematopoietic stem cells from a donor. If the transplant takes hold, the donor's stem cells set up shop in the patient's bones and begin producing a whole new, healthy blood and immune system from scratch. It's one of the more striking treatments in modern medicine, and it works precisely because everything in the immune system flows from that one small population of stem cells.

Keep this origin in mind as you move into the next chapter. Every cell you're about to meet, the neutrophils, the macrophages, the T cells, the B cells, all of them came from the same marrow and the same handful of stem cells. They branch out into wildly different jobs, but they share a common birthplace. Understanding where they come from makes it easier to see how they fit together into a single, coordinated system.

Discussion Question

The immune system has to make a hard trade-off: respond too weakly and you can't fight off real threats, but respond too strongly and you damage your own body or attack harmless things like food and beneficial bacteria. Given that both failures are dangerous, how do you think the body should "decide" where to set that balance, and what might cause the balance to drift too far in one direction over a person's lifetime?

Chapter 2: Cells of the Immune System

In the last chapter, you learned that every immune cell starts in the bone marrow and branches into different jobs. Now it's time to meet those cells individually. This is where immunology stops being abstract and gets personal, because each of these cells has its own personality, its own way of working, and its own moment to shine.

Think of what follows as introductions at a party. Some of these cells are blunt, fast, and a little reckless. Others are patient strategists. A few are trained assassins. Together they make up the workforce that keeps you alive, and knowing who's who will make everything in the rest of this book easier to follow.

Neutrophils and Phagocytosis

When bacteria break through your skin, one cell arrives before all the others: the neutrophil. It's your body's first responder, the immune equivalent of the fire truck that reaches the scene while everyone else is still getting their boots on. Neutrophils are produced in enormous numbers and stand ready in your bloodstream at all times, waiting for a call. When one comes, they can reach a wound within an hour.

To understand neutrophils, you first need to understand the process they're built around, because it's one of the oldest and most important tricks in all of immunity. That process is called phagocytosis, from the Greek words meaning "cell eating." A phagocyte is a cell that engulfs and digests other things, and phagocytosis is the act of doing it. This isn't a modern discovery. Back in 1882, a Russian zoologist named Élie Metchnikoff pushed a rose thorn into the transparent larva of a starfish and watched under his microscope as mobile cells swarmed to the injury and surrounded the intruding splinter. He'd stumbled onto the idea that certain cells actively eat invaders, and it changed biology. He shared a Nobel Prize for the work in 1908.

Here's how phagocytosis actually works, step by step. A phagocyte like a neutrophil detects a target, usually a bacterium, and flows around it, wrapping the invader in a piece of its own membrane. That membrane pinches off inside the cell, trapping the bacterium in a sealed bubble called a phagosome. Now the bacterium is caged inside the neutrophil. The cell then delivers a package of destructive enzymes and toxic chemicals into the bubble, essentially dumping acid and digestive machinery onto the trapped invader. Within minutes, the bacterium is broken down into harmless fragments. The neutrophil has eaten its enemy alive.

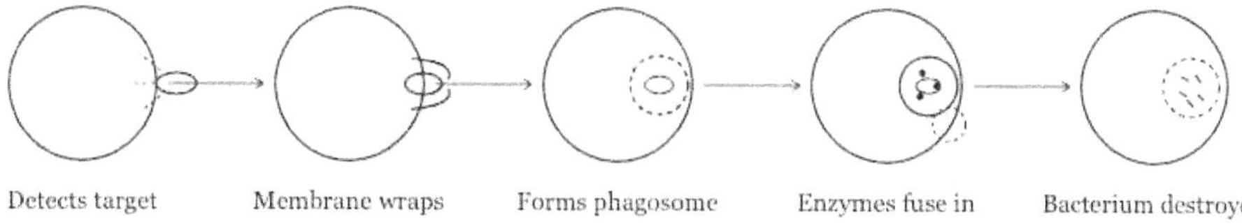

Neutrophils have a second weapon that makes this even more effective. They produce a burst of highly reactive oxygen molecules, a chemical assault sometimes called the respiratory burst, which shreds the trapped microbes from the inside. It's brutal and effective. The downside is that these chemicals don't distinguish perfectly between invader and host, so a neutrophil at full throttle causes some collateral damage to nearby tissue. That damage is part of the price of a fast, aggressive defense.

How does a neutrophil know where to go? It follows chemical trails. Damaged cells and invading microbes release signaling molecules that spread outward from the site of trouble, and neutrophils sense these gradients and swim up them toward the source, a behavior called chemotaxis. Picture following the smell of smoke to find a fire. The stronger the scent gets, the closer the neutrophil is to the problem, until it arrives right where it's needed.

Neutrophils are also famously self-sacrificing. They're short-lived by design, and most of them die in the act of fighting. When you see pus in an infected cut, you're looking at the aftermath of a battle: pus is largely made of dead neutrophils, spent invaders, and tissue debris. It's a sign your immune system showed up and fought, even if the fight got messy. The body treats neutrophils as disposable troops, cheap to produce and sent in by the millions, expecting many of them not to come back.

One of the more surprising discoveries about these cells came in 2004, when researchers found that neutrophils have a dramatic last resort. As they die, some of them fling out sticky webs made of their own DNA, studded with toxic proteins. These nets, called neutrophil extracellular traps, catch and hold bacteria in place, immobilizing invaders even after the neutrophil itself is gone. It's a genuinely startling image: a dying cell unraveling its own genetic material to snare its enemies in a trap. Scientists are still working out all the roles these nets play, including how they sometimes cause problems in diseases where they form in the wrong places.

For all their power, neutrophils have real limits. They're generalists with no memory. They respond the same way every time, they can't tell one bacterium from another in any detailed sense, and once the immediate fight is over, they leave no lasting record. They buy time and blunt the first wave of an attack, but they can't win a war on their own. That job requires the cells you'll meet next.

Macrophages and Dendritic Cells

If neutrophils are the reckless first responders, macrophages are the seasoned veterans, and dendritic cells are the intelligence officers. These two cell types are also phagocytes, so they eat invaders the same basic way neutrophils do, but they bring capabilities that go far beyond simple destruction. Both of them do something neutrophils can't: they connect the fast innate response to the slow, precise adaptive one. Without them, the two halves of your immune system would never talk to each other.

Start with macrophages. The name means "big eater," which Metchnikoff coined right alongside his work on phagocytosis, and it fits. Macrophages are large, hungry, and long-lived, sticking around in your tissues for weeks, months, or even years rather than dying off within a day. They're the cleanup crew and the neighborhood watch rolled into one. They patrol your tissues constantly, swallowing dead cells, cellular debris, and any invaders they come across, while also keeping an eye out for signs of trouble.

What makes macrophages so interesting is that they set up permanent residence throughout your body, and they take on specialized identities depending on where they live. The macrophages in your liver are called Kupffer cells, and they filter your blood as it passes through. The ones in your lungs, called alveolar macrophages, sweep up the dust, pollen, and microbes you inhale with every breath. Your brain has its own resident macrophages called microglia, which watch over your nervous system. Even your bones contain macrophage relatives, the osteoclasts, which remodel bone tissue. These are all variations on the same cell, tuned to the needs of the tissue they call home. Most of them start life as a circulating cell called a monocyte, which travels in your blood and then settles into a tissue, where it matures into a macrophage suited to that location.

Macrophages do more than eat. They're chemical communicators, releasing signaling molecules called cytokines that shape the immune response happening around them. When a macrophage encounters a serious threat, it pumps out cytokines that recruit more immune cells, ramp up inflammation, and raise the alarm throughout the neighborhood. Later, once the threat is handled, macrophages switch modes and release different signals that calm the inflammation down and help repair the damaged tissue. The same cell that helped start the fight helps clean up afterward and rebuild. This flexibility, the ability to shift from aggressive attacker to gentle healer depending on what the situation calls for, is one of the reasons macrophages are so central to how tissues stay healthy.

Now for the dendritic cell, which might be the most important messenger in your entire immune system. Dendritic cells get their name from their shape: they have long, branching arms, called dendrites, that reach out in all directions like the branches of a tree. Those arms give them enormous surface area for catching invaders. A researcher named Ralph Steinman discovered these cells in 1973, and it took decades for the field to fully appreciate what he'd found. He was awarded a Nobel Prize for the work in 2011. In one of the more poignant moments in the history of science, Steinman died just three days before the prize was announced, and the Nobel committee, in an unusual decision, let the award stand.

Here's what makes dendritic cells so special. They station themselves in your skin, your gut lining, your lungs, and other places where invaders are likely to enter. There, they act as sentinels, constantly sampling their surroundings and capturing any microbes they encounter through phagocytosis. But instead of simply destroying what they catch and moving on, a dendritic cell does something clever. It breaks the invader into pieces and holds onto samples of it. Then it leaves the front lines entirely, traveling through the lymphatic system to the nearest lymph node, carrying its captured evidence with it. In the lymph node, packed with adaptive immune cells waiting for information, the dendritic cell shows off the pieces of the invader it collected. (This is illustrated at the end of this section.) This act, called antigen presentation, is the handoff that launches the adaptive immune response. The dendritic cell is essentially a courier, running captured intelligence from the battlefield back to headquarters so the specialists can build a targeted response.

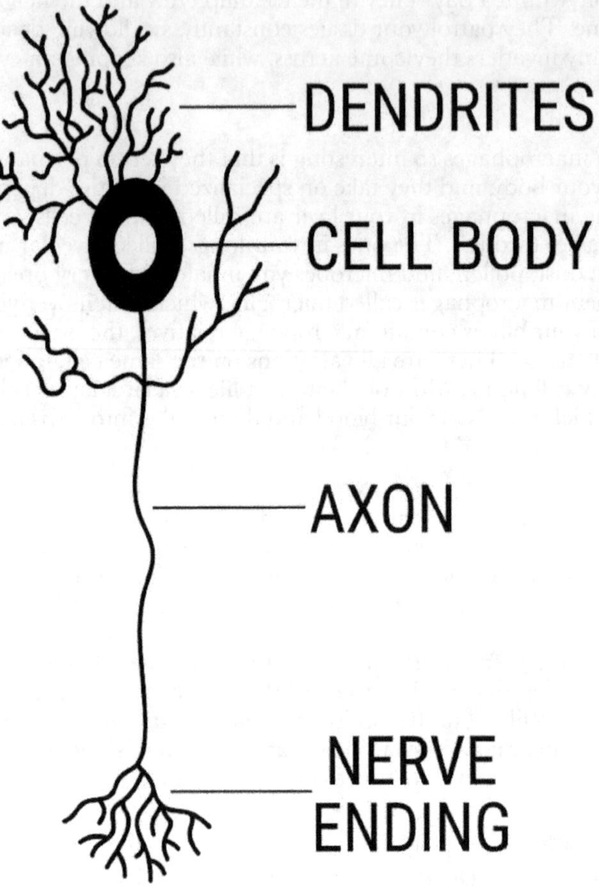

This is the bridge you read about in the last chapter, made concrete. The innate system, through the dendritic cell, gathers information and delivers it to the adaptive system. No dendritic cells, no handoff, and the precise, memory-based defense of adaptive immunity would never get activated. That's why immunologists often call dendritic cells the most powerful antigen-presenting cells in the body, and why they sit at the crossroads of the whole immune response.

Let's compare the three phagocytes you've now met, since they're easy to mix up.

Cell	Lifespan	Primary Job	Special Trait
Neutrophil	Short, often a day or less	First responder; rapid killing	Casts DNA nets; dies fighting

Macrophage	Long, weeks to years	Cleanup, defense, tissue repair	Takes on local identities in each organ
Dendritic cell	Variable	Antigen presentation	Bridges innate and adaptive immunity

All three eat invaders. But they eat for different reasons and with different follow-through. The neutrophil eats to kill fast. The macrophage eats to clean, defend, and heal over the long haul. The dendritic cell eats to learn, and then it carries what it learned to the cells that can act on it.

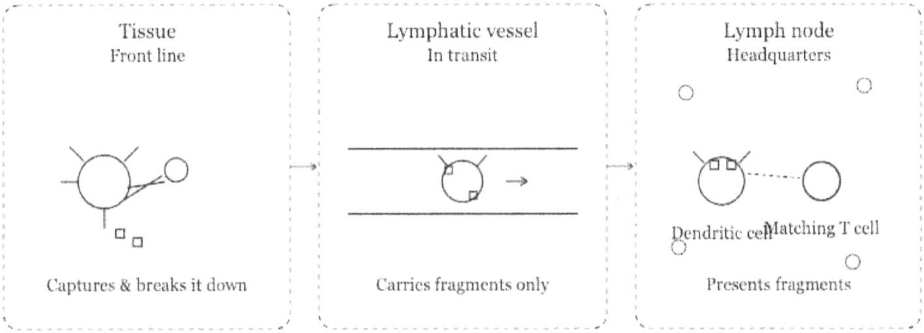

Information travels, not the whole pathogen

T Lymphocytes and Their Subtypes

Now we reach the heart of adaptive immunity. T lymphocytes, or T cells, are among the most sophisticated cells in your body, and they run much of the targeted, intelligent defense that keeps chronic and serious infections in check. The "T" stands for thymus, the organ where these cells go to be trained, as you learned in the last chapter. Only the T cells that pass their training, learning to attack foreign threats without turning on your own body, ever make it into circulation.

What sets T cells apart is how they recognize their targets. Each T cell carries a receptor on its surface, called the T cell receptor, that's tuned to recognize one specific molecular shape and nothing else. Your body produces an astonishing variety of these receptors, so that between all your T cells, you have specialists ready for almost any invader imaginable, including ones that have never existed before. But there's a twist. A T cell can't recognize a free-floating invader the way an antibody can. It only recognizes its target when that target is displayed on the surface of another cell, presented in the grip of those MHC molecules you met in the last chapter. In other words, T cells read the ID tags and displayed samples that cells carry on their surface. This requirement, that T cells see antigens only when they're presented properly, is one of the defining features of how they work.

T cells come in several types, each with a different mission. The two you'll hear about most are helper T cells and cytotoxic T cells, and the difference between them starts with which display molecule they read.

Helper T cells are the coordinators, the field generals of the immune response. Scientists identify them by a surface marker called CD4, so you'll often see them written as CD4 cells. Helper T cells don't kill anything directly. Instead, they read the samples presented by antigen-presenting cells like dendritic cells, and once activated, they release a flood of cytokines that direct the rest of the immune system. They tell B cells to start making antibodies. They call in and activate other cells. They amp up the killers. Almost nothing in the adaptive response happens well without helper T cells organizing it, which is exactly why they're such a devastating target when they're lost. The virus that causes AIDS, HIV, specifically infects and destroys helper T cells, and as their numbers fall, the entire immune response falls apart with them. We'll return to that in a later chapter, but it's worth seeing here just how central these coordinators are.

Cytotoxic T cells are the assassins. They carry a different surface marker, CD8, so you'll see them called CD8 cells. Their job is to find and kill your own cells when those cells have been compromised, usually because a virus has infected them or because they've turned cancerous. This raises a question: how does a cytotoxic T cell know a cell is infected from the outside? The answer is elegant. Every cell in your body constantly displays samples of what's being made inside it, held up on MHC molecules for inspection. A healthy cell displays only normal, self proteins. A virus-infected cell, forced to manufacture viral proteins, ends up displaying bits of the virus on its surface without meaning to. Cytotoxic T cells patrol past, read these displays, and when they spot viral or abnormal fragments, they know the cell has been taken over. They then kill it, sacrificing one infected cell to stop the virus from spreading. They do this by punching holes in the target cell and triggering it to self-destruct, a controlled demolition that we'll look at more closely with the natural killer cells shortly.

There's a third type that deserves attention: the regulatory T cell, often shortened to Treg. If helper T cells are the accelerator, regulatory T cells are the brakes. Their job is to keep the immune response under control, to shut it down when it's no longer needed, and to prevent it from attacking your own body. They enforce the tolerance that stops your immune system from causing autoimmune disease. When regulatory T cells fail or are too few, the immune system can slip into attacking healthy tissue. When they're too active, they can suppress responses that should be happening, sometimes even shielding tumors from attack. They walk a fine line, and your health depends on them walking it well.

Finally, there are memory T cells, the reason adaptive immunity has such a long memory. After an infection is defeated, most of the T cells that fought it die off, their work done. But a small number survive and persist for years or even decades, quietly remembering the specific invader they were built to fight. If that same invader ever returns, these memory T cells recognize it at once and mount a response far faster and stronger than the first time. Memory T cells are a large part of why you don't get certain diseases twice, and they're one of the outcomes that vaccines are designed to produce.

Here's a summary of the T cell family.

T Cell Type	Marker	Role
Helper T cell	CD4	Coordinates the immune response; directs other cells
Cytotoxic T cell	CD8	Kills infected and cancerous cells directly
Regulatory T cell	CD4 (specialized)	Suppresses and controls the response; maintains tolerance
Memory T cell	Varies	Remembers past invaders for a faster future response

Take a moment with this table, because these four roles come up constantly in immunology. A great deal of what happens in infection, vaccination, autoimmune disease, and cancer treatment comes down to these cells: which ones get activated, which ones are missing, and which ones are working too hard or not hard enough.

B Lymphocytes and Antibody Production

The other great cell of adaptive immunity is the B lymphocyte, or B cell. If T cells run the coordination and the direct killing, B cells run the antibody program, and antibodies are among the most versatile weapons your body makes. B cells are the reason your blood contains a supply of custom-built proteins ready to latch onto specific invaders.

The letter "B" has a curious origin. It comes from an organ in birds called the bursa of Fabricius, where these cells were first found to mature. In humans and other mammals, we don't have that organ, but conveniently, B cells mature in our bone marrow, so the letter still works as a memory aid. Like T cells, B cells go through a screening process during their development, weeding out any that would attack the body's own tissues, so that only safe, useful cells enter circulation.

Each B cell, like each T cell, is a specialist tuned to recognize one particular target. A B cell carries its receptor on its surface too, but here's a way B cells differ from T cells that matters enormously: a B cell's receptor is essentially an antibody, anchored to the cell membrane. And unlike T cells, which can only see antigens presented on other cells, B cells can recognize invaders directly, latching onto free-floating targets in your blood and tissues. This gives them a different reach than T cells have.

When a B cell encounters its specific target and gets the proper go-ahead signals, usually with help from those helper T cells you just met, it springs into action and transforms. It divides rapidly and its offspring split into two different careers. Most of them become plasma cells, which are the antibody factories, and a smaller number become memory B cells, which stand by for the future.

Plasma cells are production machines, and their output is genuinely hard to fathom. A single plasma cell can manufacture and release something like two thousand antibody molecules every second. Multiply that across the many plasma cells generated during an infection, and your body floods itself with billions upon billions of these targeting proteins, all shaped to grab one specific invader. Each plasma cell pours out antibodies continuously until the infection is controlled, blanketing your bloodstream with weapons aimed at the enemy.

So what do all these antibodies actually do once they're made? They don't kill invaders on their own the way a cytotoxic T cell does. Instead, antibodies work by binding tightly to their targets, and that binding accomplishes several things at once. An antibody can coat an invader and block it from attaching to your cells, a trick called neutralization that stops many viruses cold. Antibodies can also tag invaders for destruction, coating a bacterium so that phagocytes recognize it instantly and gobble it up, a process that turns a hard-to-catch microbe into an easy meal. And antibodies can flag targets for other parts of the immune system to finish off. In short, antibodies mark, block, and disable, then hand the actual killing off to other players. The next few chapters go much deeper into how antibodies are built and the different forms they take, so for now, hold onto the core idea: B cells become plasma cells, and plasma cells make antibodies by the billions.

The memory B cells matter just as much as the antibodies. After the infection clears and most plasma cells die off, these memory cells linger for years, holding onto the blueprint for the exact antibody that worked. If the same invader returns, memory B cells reactivate quickly, spinning up fresh plasma cells and antibody production far faster than the first time around. This is the antibody side of immune memory, and it works hand in hand with the memory T cells to give you lasting protection. When a vaccine gives you immunity that lasts for years, memory B cells are a big part of the reason why.

Natural Killer Cells

The last cell in this chapter is a bit of a hybrid, and it fills a gap the other cells leave open. Natural killer cells, usually shortened to NK cells, come from the same lymphoid family as T and B cells, but they behave like part of the innate system: they're fast, ready from the start, and they don't need to learn about a threat before they act. That combination makes them unusual and useful.

The name tells you their job. They're natural born killers, able to destroy target cells right away without any prior sensitization or training against a specific enemy. A cytotoxic T cell has to be activated and directed at a particular target before it can kill. An NK cell doesn't wait for that. It patrols your body and can strike compromised cells immediately, which makes it especially useful in the early hours of a viral infection, before the slower adaptive response has geared up.

What makes NK cells clever is how they choose their targets, and it's the mirror image of how cytotoxic T cells work. Remember that healthy cells display MHC molecules on their surface, the ID tags showing they belong. NK cells check for exactly these tags, but they read them in reverse. A cell showing proper MHC tags gets left alone. A cell that's missing its tags gets killed. This is sometimes called "missing self" recognition, and it solves a sneaky problem. Some viruses and many cancer cells have learned to hide their MHC molecules precisely to escape detection by cytotoxic T cells, which need those tags to spot infected cells. By pulling down their ID tags, these cells slip past the T cell patrol. But that same trick makes them stand out to NK cells, which are watching for cells that have gone dark. The two systems cover each other's blind spots. A cell can't hide from both at once: keep your ID tags up and the T cells might catch you, take them down and the NK cells will.

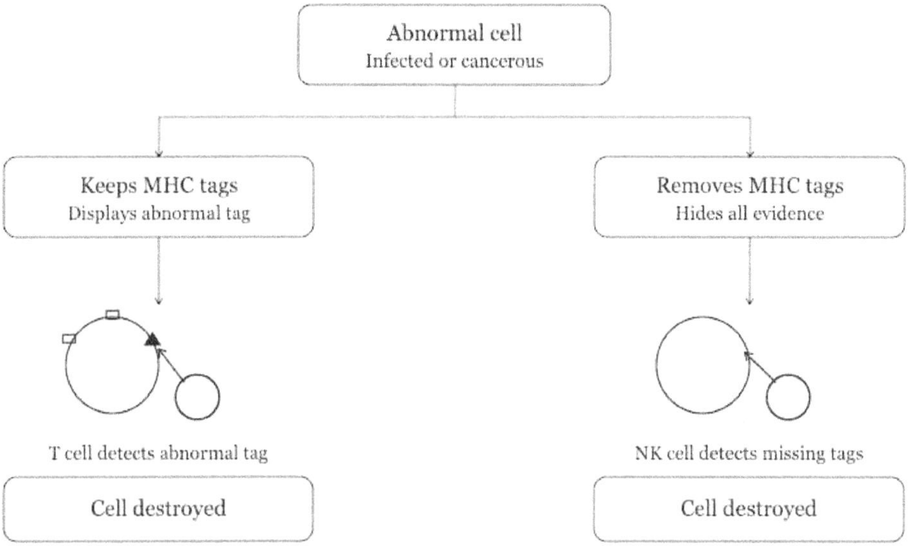

When an NK cell decides to kill, it uses the same method as a cytotoxic T cell, which is worth understanding because it's how your body eliminates its own compromised cells. The killer cell releases two kinds of molecules onto the target. The first, called perforin, punches tiny holes in the target cell's outer membrane. Through those holes, the killer delivers the second kind, enzymes called granzymes, which switch on the target cell's self-destruct program.

Rather than bursting the cell messily, this triggers the target to die in a clean, controlled way, packaging itself up for safe disposal. It's an efficient, tidy execution that stops a virus from spreading without spilling infectious contents everywhere.

NK cells do more than kill, too. Like macrophages, they release cytokines, especially one called interferon-gamma, that shape the wider immune response and help activate other cells. So they're not just assassins; they also send signals that strengthen the overall defense. Their combined ability to kill compromised cells on sight and to sound chemical alarms makes them important against both viruses and developing cancers, standing guard in the window of time before adaptive immunity arrives.

Here's a quick comparison between NK cells and cytotoxic T cells, since they share a killing method but differ in how they pick targets.

Feature	Natural Killer Cell	Cytotoxic T Cell
Immune branch	Innate	Adaptive
Needs prior training	No, acts immediately	Yes, must be activated first
Target recognition	Kills cells missing MHC tags	Kills cells displaying foreign fragments on MHC
Killing method	Perforin and granzymes	Perforin and granzymes
Memory	Limited	Yes, forms memory cells

You've now met the main cast of the immune system: the fast-eating neutrophils, the versatile macrophages, the messenger dendritic cells, the coordinating and killing T cells, the antibody-making B cells, and the ever-ready natural killer cells. Each one came from the same bone marrow, and each one does a different job. In the chapters ahead, you'll see how they work together in real responses, starting with the fast innate defenses that meet invaders first.

Discussion Question

Natural killer cells and cytotoxic T cells destroy compromised cells using the same weapons, yet one belongs to the innate system and acts instantly while the other belongs to the adaptive system and needs training first. Why might it benefit the body to maintain two separate cell types that kill the same way but operate on such different timelines, rather than relying on just one?

Chapter 3: The Innate Immune Response

Here's something most people never consider: the vast majority of microbes that try to invade you never make it past the front door. They don't get killed by a heroic immune cell. They simply can't get in. Your body's first line of defense isn't a battle at all. It's a wall.

The innate immune response is everything your body does to stop and fight invaders without needing to learn about them first. It's fast, it's always on, and it handles the overwhelming majority of threats you encounter every day. In this chapter, we start at the outermost layer, the barriers that keep invaders out entirely, and work inward to the alarm systems, the protein cascades, and the coordinated response called inflammation that together make up your innate defenses.

Physical and Chemical Barriers

Before your immune system ever has to fight, it tries to make sure there's nothing to fight. Your barriers, both physical and chemical, block, trap, and destroy invaders before they can establish a foothold. These defenses are so effective and so constant that you never notice them working, but they stop far more infections than any of the dramatic cellular battles you read about in the last chapter.

Your skin is the obvious starting point, and it's a better fortress than it looks. The outer layer, the part you can see and touch, is made of flattened, dead cells packed with a tough protein called keratin, the same material in your hair and nails. Living cells are vulnerable. Dead, keratin-filled ones are not. This outer sheet is dry, and most microbes need moisture to grow, so the very dryness of your skin surface makes it a hostile place to live. On top of that, your skin sheds constantly. You lose millions of dead skin cells every day, and any bacteria that settled on the surface get carried off with them. It's like trying to build a house on a floor that's being swept out from under you around the clock.

Your skin fights chemically too. Oil glands coat the surface with sebum, which contains fatty acids that many microbes can't tolerate. The combined effect of sweat, sebum, and skin secretions keeps the surface slightly acidic, at a pH of roughly five, well below the neutral range most bacteria prefer. This so-called acid mantle discourages the growth of harmful microbes while your own resident bacteria, the ones adapted to living on you, get along fine.

Skin covers your outside, but your inside surfaces need protection too, and they can't use the same dry, dead-cell approach. Your airways, your digestive tract, and your urogenital tract all need to stay moist and living to do their jobs, which makes them more inviting to microbes. These surfaces are lined with mucous membranes, and their main weapon is mucus itself. That sticky fluid you produce in your nose, throat, and gut traps incoming microbes and particles like flypaper, holding them so they can't reach the cells underneath.

Trapping invaders is only useful if you can then get rid of them, and your body has a clever conveyor belt for exactly that. The cells lining your airways are covered in tiny hair-like structures called cilia, which beat in coordinated waves, thousands of times a minute, pushing the mucus and everything stuck in it upward and out of your lungs. This mucociliary escalator carries trapped debris up to your throat, where you swallow it without noticing, sending it down to your stomach acid to be destroyed. When you're sick and coughing up phlegm, you're seeing this system working overtime. It's also part of why smoking is so damaging: the chemicals in cigarette smoke paralyze and destroy cilia, shutting down the escalator and leaving the lungs far more open to infection.

Beyond mucus, your body deploys an arsenal of antimicrobial chemicals. Your tears, saliva, and mucus all contain an enzyme called lysozyme, which chops apart the cell walls of many bacteria, breaking them open. A Scottish scientist named Alexander Fleming discovered lysozyme in 1922, several years before he stumbled onto penicillin, and he found it by testing his own nasal mucus during a cold. Your stomach adds a far blunter weapon: gastric acid so strong, at a pH between one and two, that it destroys most microbes you swallow along with your food. Your body also produces small proteins called defensins, natural antibiotics that punch holes in microbial membranes across your skin and mucous surfaces.

Some of your defenses are simply mechanical flushing. Tears wash your eyes clean. The steady flow of urine flushes bacteria out of your urinary tract, which is one reason drinking plenty of water helps prevent urinary infections. Coughing and sneezing forcibly expel invaders from your airways. Even the flow of saliva keeps rinsing your mouth. These aren't sophisticated, but constant washing keeps microbes from settling and building up.

Finally, you have living defenders on your surfaces that aren't part of your immune system at all: your own resident bacteria. The trillions of harmless microbes on your skin and in your gut occupy space, consume nutrients, and produce substances that keep dangerous newcomers from gaining a foothold. They crowd out invaders through sheer competition. This is why antibiotics, which kill off your helpful bacteria along with the harmful ones, can leave you vulnerable to opportunistic infections. Wipe out the good tenants, and the bad ones move into the empty rooms.

Together, these barriers form a defense that asks nothing of the rest of your immune system. No cells have to fight, no alarms have to sound, and you never feel a thing. But barriers aren't perfect. Cuts, scrapes, burns, and the ordinary wear of daily life create openings, and some microbes are simply good at getting through. When something does break in, your body needs a way to detect it fast, which brings us to how innate immunity actually recognizes a threat.

Pattern Recognition Receptors

Once an invader breaches your barriers, your innate immune cells face a hard question, the same one from Chapter 1 in a new form: how do you recognize a threat you've never seen, instantly, without any time to learn about it? The adaptive system answers this by building custom receptors over days. The innate system can't wait that long. It needs to identify danger in seconds. Its solution is one of the most elegant ideas in all of immunology.

Innate immune cells don't try to recognize each specific microbe. Instead, they recognize broad molecular signatures that whole classes of microbes share, features that are common to many bacteria or many viruses but absent from your own cells. These telltale molecular fingerprints are called pathogen-associated molecular patterns, usually shortened to PAMPs. The word "pattern" is the important part. Rather than memorizing every individual criminal, innate cells learn to spot the uniforms that entire categories of criminals wear.

What kinds of patterns are we talking about? Consider a few real examples.

A molecule called lipopolysaccharide, or LPS, coats the outer surface of a huge group of bacteria and appears nowhere in the human body, so its presence is a reliable signal that certain bacteria have invaded. Flagellin, the protein that makes up the whip-like tails many bacteria use to swim, is another giveaway. Double-stranded RNA is a hallmark of many viruses replicating inside your cells, since your own cells don't normally make it. Peptidoglycan, a building block of bacterial cell walls, is yet another. Each of these molecules is essential to the microbe and shared across many species, which makes it a dependable target. A microbe can't easily abandon its own cell wall or its own genetic material just to hide, so these patterns are hard to fake and hard to lose.

Your cells detect these patterns using sensors called pattern recognition receptors, or PRRs. These are proteins, built into your innate immune cells from birth according to instructions in your DNA, each one shaped to lock onto a particular molecular pattern. When a PRR binds its target pattern, it sets off an alarm inside the cell, triggering the cell to spring into action: to start eating, to release cytokines, to call for reinforcements, or to ramp up inflammation. The recognition and the response are wired together.

The most famous family of these sensors is the Toll-like receptors, and their discovery story is a good one. In the 1980s, a gene called Toll was found in fruit flies, where it helped control how the embryo developed its body plan. It seemed to have nothing to do with immunity. But in 1996, a French scientist named Jules Hoffmann discovered that the same Toll gene was also essential for the fruit fly's defense against fungal infection. Flies with a broken Toll gene got overwhelmed by fungus and died. Soon after, an American researcher named Bruce Beutler found that mammals, including humans, have their own versions of these receptors, which he showed were the sensors that detect bacterial LPS. The two of them shared a Nobel Prize in 2011, alongside Ralph Steinman of dendritic cell fame, for uncovering how the innate immune system recognizes invaders. Humans turn out to have around ten different Toll-like receptors, and between them they cover a wide range of microbial patterns.

Where these receptors sit inside the cell is itself a clever piece of design. Some Toll-like receptors sit on the outer surface of the cell, facing the outside world, where they catch the surface features of bacteria and fungi passing by. Others are placed inside the cell, on the membranes of internal compartments, where they wait to detect the genetic material of viruses that have been swallowed and broken open. This makes sense: bacterial cell-wall components are found outside, so the sensors for them face outward, while viral nucleic acids show up inside the cell, so the sensors for them face inward. The receptors are positioned exactly where their targets are likely to appear.

Toll-like receptors aren't the only pattern sensors you have. Another family, the NOD-like receptors, patrols the inside of your cells for signs of bacterial invasion in the cytoplasm. A related family, the RIG-I-like receptors, hunts specifically for viral RNA that has slipped into the cell's interior. Together, these families give you sensors covering essentially every location an invader might try to hide: on the cell surface, inside internal compartments, and free within the cell's main body. Wherever a microbe goes, something is watching for it.

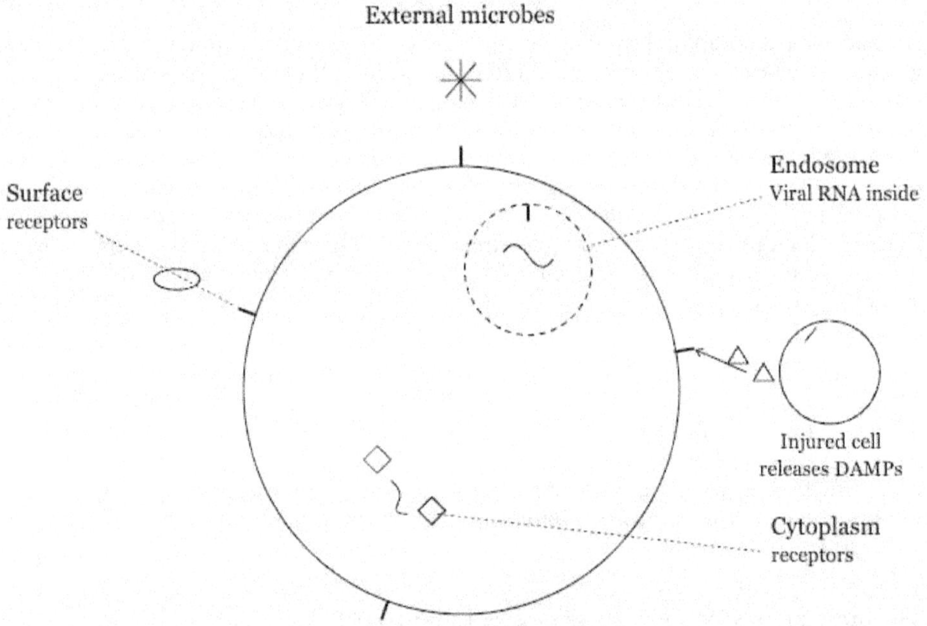

Wherever a microbe goes, something is watching

One particular assembly deserves a mention because it comes up often in modern medicine. Certain NOD-like receptors can gather together with other proteins into a large complex called the inflammasome. When it detects the right danger signals, the inflammasome activates and triggers the release of powerful inflammatory cytokines, ramping up the response fast. It's essentially a molecular trigger that turns detection into full-blown inflammation, and it's involved in a range of diseases when it fires inappropriately.

Here's what makes this whole system so different from adaptive immunity, and it's worth pausing on. Your pattern recognition receptors are limited in number, maybe a few dozen types in total, and they're the same in you as in the next person, encoded directly in the genome you were born with. They can't learn, and they can't adapt to a brand-new pattern. But they don't need to. By targeting the deep, unchanging features that microbes can't do without, a small fixed set of sensors covers a vast range of threats. The adaptive system, by contrast, can build billions of different receptors and recognize almost anything, but only after days of preparation. Innate recognition trades that limitless flexibility for speed and reliability. It knows fewer things, but it knows them instantly and it never forgets, because the knowledge is built into your genes.

There's one more category these sensors detect, and it connects back to an idea from the first chapter. Pattern recognition receptors don't only sense invaders. They also detect signals released by your own damaged and dying cells, molecules that normally stay locked inside a cell and only spill out when the cell is injured. These distress signals are called damage-associated molecular patterns, or DAMPs. When your tissue is hurt, even without any microbe present, DAMPs leak out and trigger the same alarm receptors, kicking off an immune response to the injury. This is the "danger" that the danger model described: your innate system watches both for foreign patterns and for signs that your own cells are under attack, and either one is enough to sound the alarm.

The Complement System

Not all of your innate defenses are cells. Some of the most effective ones are proteins floating in your blood, ready to activate the instant they encounter an invader. The complement system is the best example, and it's one of the more underappreciated parts of your immune arsenal. It's a set of more than thirty proteins, made mostly by your liver, that circulate constantly in your bloodstream in an inactive, dormant state, waiting.

The name is a small window into the history of immunology. When scientists first studied it over a century ago, they found it "complemented," or completed, the work of antibodies in destroying bacteria, so they called it complement. The Belgian scientist Jules Bordet did much of the early work and won a Nobel Prize for it in 1919. The individual proteins were numbered as they were discovered, from C1 through C9, which unfortunately means the numbers don't follow the order in which the proteins actually act. It's a quirk you just have to accept: the naming reflects the sequence of discovery, not the sequence of the reaction.

Complement works as a cascade, and understanding that word is the whole trick. A cascade is a chain reaction where one activated protein activates the next, which activates the next, and so on, amplifying enormously at each step. One triggering event sets off a domino chain, and because each activated protein can switch on many copies of the next, the response grows fast. A tiny initial signal becomes a massive coordinated attack within moments. This amplification lets the system respond powerfully to even a small number of invaders.

There are three different ways to set off the cascade, and all three funnel into the same central point. The classical pathway is triggered when antibodies bind to an invader, linking complement to the adaptive immune system and giving antibodies a way to summon this destructive machinery. The lectin pathway starts when a blood protein recognizes certain sugars found on the surface of many microbes but not on your own cells, catching invaders directly. The alternative pathway is the simplest:

complement proteins spontaneously activate at a low level all the time, and if they land on a microbe's surface rather than your own protected cells, the reaction takes hold and amplifies. Three different triggers, three different starting points, but all three converge on a single protein, C3, which sits at the heart of the system. Once C3 is activated, the real work begins.

Activated complement produces three main outcomes, and each one attacks invaders in a different way. The first is tagging invaders for destruction. Fragments of complement, especially one called C3b, coat the surface of a microbe, and phagocytes carry receptors that grab onto this coating. A bacterium slick with complement becomes far easier for a neutrophil or macrophage to catch and eat, the same phagocytosis you read about in the last chapter, now made dramatically more efficient. This coating-for-easier-eating is called opsonization, from a Greek word that roughly means "to prepare for eating." Complement essentially seasons the invader to make it a more appetizing target.

The second outcome is stirring up inflammation. Some of the small fragments produced during the cascade, particularly ones called C3a and C5a, act as chemical alarm signals. They spread through the tissue and summon immune cells to the site, guiding neutrophils and other cells toward the invasion by chemotaxis. They also make blood vessels leakier and trigger the release of inflammatory molecules, jump-starting the broader inflammatory response we'll get to next. In this way, complement doesn't just attack invaders directly; it also recruits the rest of the immune system to the fight.

The third outcome is the most dramatic: complement can kill microbes outright by tearing holes in them. In the final steps of the cascade, several complement proteins assemble together into a ring-shaped structure called the membrane attack complex. This complex embeds itself in the outer membrane of a target microbe and punches a pore straight through it. With a hole in its membrane, the microbe can't maintain its internal balance; fluid rushes in, its contents leak out, and it bursts and dies. This is complement acting as a direct executioner, destroying certain bacteria without any help from immune cells at all.

There's an obvious danger lurking in all this. If complement can punch holes in bacterial membranes and tag surfaces for destruction, what stops it from attacking your own cells, which are sitting right there in the same bloodstream? The answer is that your cells carry protective proteins on their surfaces that shut complement down before it can hurt them, regulatory molecules that act as a safety switch, telling complement "this one's friendly, stand down." Microbes lack these protections, so complement destroys them while sparing you. When this regulation fails, complement can turn on the body's own tissues and cause disease, which is exactly what happens in certain serious blood and kidney disorders. The system is destructive by design, so keeping it pointed only at invaders is essential.

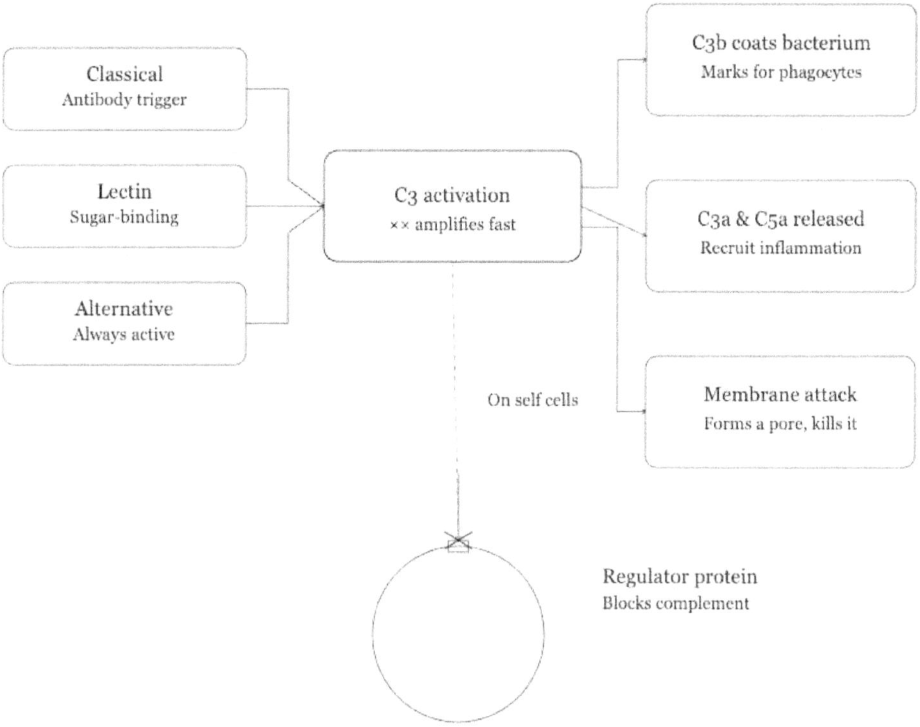

Complement destroys microbes without help, but self stays protected

Here's a summary of what complement accomplishes once it's activated.

Complement Function	How It Works	Result
Opsonization	C3b coats the microbe's surface	Phagocytes catch and eat it more easily
Inflammation	C3a and C5a summon immune cells	More defenders arrive; vessels become leaky
Membrane attack complex	Proteins form a pore in the microbe	The microbe bursts and dies directly

Complement ties together several themes you've now seen. It works with antibodies through one of its pathways, it makes phagocytosis more effective, and it drives inflammation. It's a bridge between many parts of your defenses, and it does its work in seconds, entirely through proteins already circulating in your blood.

Inflammation as a Defense Mechanism

Everything in this chapter so far comes together in inflammation, the coordinated response your body mounts at the site of an injury or infection. You've felt it many times. The redness and swelling around an infected cut, the heat and tenderness of a sprained ankle, the puffiness of an insect bite, these are all inflammation. It feels like a problem, but it's actually your immune system organizing its defense, and understanding what it's doing changes how you see every ache and swelling you've ever had.

Inflammation is so recognizable that physicians described it nearly two thousand years ago. Around the first century AD, a Roman writer named Celsus laid out four signs of inflammation that doctors still teach today: redness, heat, swelling, and pain. In Latin, they're rubor, calor, tumor, and dolor. Later, the Greek physician Galen added a fifth: loss of function, the way an inflamed joint or wound stops working normally while it heals. These five signs have described inflammation accurately since antiquity, long before anyone understood the cells and molecules behind them.

So what's actually happening beneath those signs? Each one is the visible result of a specific defensive action, and once you connect the sign to its cause, inflammation stops looking like damage and starts looking like strategy.

The redness and heat come from the same source: increased blood flow. When tissue is injured or invaded, the small blood vessels in the area widen, a process called vasodilation, which floods the site with extra blood. More blood means more redness and more warmth, which is why an inflamed area looks pink and feels hot to the touch. But the point isn't the color or the temperature. The point is what the blood carries. That rush of blood delivers immune cells and defensive proteins, including the complement you just read about, straight to where they're needed. The vessels widen precisely so more defenders can arrive faster.

The swelling comes from a change in the blood vessels themselves. During inflammation, the vessel walls become leaky, opening tiny gaps that let fluid, proteins, and immune cells pass out of the bloodstream and into the surrounding tissue. Normally, blood vessels hold their contents in. During inflammation, they deliberately loosen up so that defenders can exit the bloodstream and reach the actual site of trouble in the tissue, where the invaders are. The fluid that leaks out along with them causes the visible swelling. That puffiness is the cost of opening the vessels so immune cells can get out of the highway and onto the local streets where the fight is happening.

Much of this vascular change is triggered by a chemical you've probably heard of: histamine. Histamine is released by mast cells, immune cells stationed throughout your tissues, especially near blood vessels and body surfaces. Mast cells act like tripwires. When they detect injury or invasion, they dump their stores of histamine, which immediately makes nearby vessels widen and turn leaky, kicking off the redness and swelling. This is also why antihistamine medications reduce the swelling, redness, and itch of allergic reactions and insect bites: they block histamine's effects on your blood vessels. When you take an antihistamine for a bee sting or hay fever, you're interrupting this exact step of the inflammatory process.

The pain of inflammation has its own purpose, even though it's the part we like least. Some of the chemicals released during inflammation directly stimulate the nerve endings in the area, producing tenderness and pain. Swelling adds to it by pressing on those nerves. It's unpleasant, but pain serves a real function: it forces you to protect the injured area. A throbbing, tender ankle makes you stop walking on it and rest, giving the tissue a chance to heal. A sore, inflamed wound makes you guard it from further harm. The discomfort is a signal your body uses to change your behavior in ways that help you recover.

The loss of function that Galen noted ties into this. An inflamed, swollen, painful body part simply doesn't work as well, and while that's frustrating, it enforces the rest that healing requires. The temporary shutdown protects the area during its most vulnerable period.

The whole process is orchestrated by chemical messengers, the cytokines you met in the last chapter. When innate cells detect an invader through their pattern recognition receptors, they release a burst of these signaling molecules, and several of them drive inflammation specifically. They tell blood vessels to widen and leak, they summon more immune cells to the area, and they crank up the whole response. Some cytokines even travel through the bloodstream to affect your entire body, which is why a bad infection makes you feel feverish, tired, and achy all over. That whole-body feeling of being sick is inflammation gone systemic, driven by cytokines circulating far beyond the original site of infection. We'll look at fever specifically in a later chapter, since your body raises its temperature on purpose as part of this response.

There's a hidden step in inflammation that matters as much as starting it: turning it off. Inflammation is destructive by nature. The same aggressive cells and chemicals that kill invaders also damage your own tissue, and if inflammation ran unchecked, it would cause serious harm. So your body has mechanisms to resolve inflammation once the threat is handled, dialing the response back down, clearing away the debris, and shifting into tissue repair. This is where those versatile macrophages come back in, switching from their attacking mode into a healing mode that rebuilds the damaged area. A healthy inflammatory response has a clear beginning, middle, and end. It flares up to fight, then shuts down and repairs.

This brings up the difference between two kinds of inflammation that has enormous consequences for health. Acute inflammation is the short-term response we've been describing: it starts fast, does its job, and resolves within days. It's helpful, even essential, and you couldn't survive without it. Chronic inflammation is what happens when the response never fully shuts off and smolders on for months or years. Instead of fighting a specific threat and resolving, chronic inflammation keeps low-grade immune activity running in tissues that should be at rest, slowly damaging them over time. This lingering, unresolved inflammation is now linked to a long list of serious conditions, and we'll come back to it in a later chapter. For now, the takeaway is that inflammation is a double-edged sword: indispensable in the short term, damaging when it won't switch off.

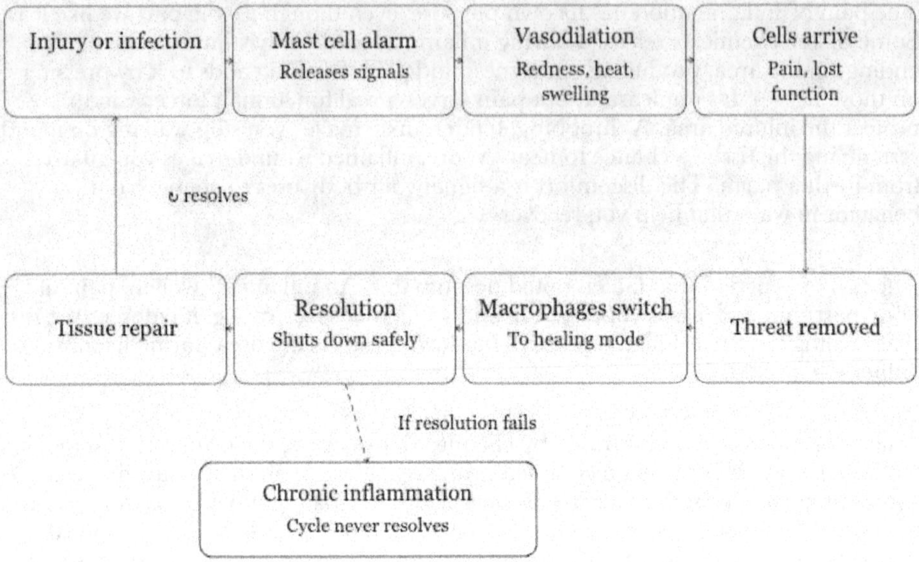

Symptoms are visible signs of each stage, not separate events

Here's a summary connecting each sign of inflammation to the defensive action behind it.

Sign of Inflammation	**Underlying Cause**	**Defensive Purpose**
Redness (rubor)	Blood vessels widen; more blood flow	Delivers immune cells and proteins to the site
Heat (calor)	Increased blood flow to the area	Speeds immune activity; may slow microbes
Swelling (tumor)	Vessels leak fluid into the tissue	Lets defenders exit blood and reach the invader
Pain (dolor)	Chemicals and swelling irritate nerves	Forces you to protect and rest the area
Loss of function	Combined swelling and pain	Enforces rest so the tissue can heal

Step back and look at what inflammation really is. It's not a malfunction or a sign that something has gone wrong with your body. It's a carefully coordinated response that brings blood, immune cells, and defensive proteins to exactly where they're needed, opens the vessels so those defenders can reach the invaders, and produces sensations that make you protect the healing tissue. Every uncomfortable part of it has a job. The next time you have a swollen, tender, warm cut, you'll know you're watching your innate immune system do precisely what it evolved to do.

Discussion Question

Inflammation causes real damage to your own tissue as a side effect of fighting invaders, and the body accepts that damage as the price of a fast defense. Given that chronic, unresolved inflammation drives so many serious diseases, why do you think evolution favored such an aggressive, self-harming response rather than a gentler one, and what does that trade-off suggest about the environments our immune systems were shaped to handle?

Chapter 4: The Adaptive Immune Response

Why does a new illness knock you flat for a week, while the second time you're exposed to the same thing you might not even notice? The answer is the subject of this entire chapter. When a truly new invader gets past your barriers and your innate defenses, your body doesn't already have the right weapon on hand. It has to find it, build it, and manufacture it in bulk, and all of that takes time, usually the better part of a week. This is the adaptive immune response, and it's the slower, smarter, more precise half of your immune system.

What makes it worth the wait is what it leaves behind. The innate response fights hard and forgets. The adaptive response fights, wins, and remembers, so that the next encounter is faster and easier. In this chapter, we'll walk through how it works from start to finish: how an invader gets introduced to the adaptive system, how the two great cell types get activated and specialized, what the antibodies they produce actually do, and how the whole thing lays down the memory that protects you for years.

Antigen Presentation

Everything in the adaptive response starts with recognizing a specific target, so let's nail down what that target is. An antigen is any molecule your immune system can recognize and respond to. The word is a squashed-together version of "antibody generator," meaning a substance that provokes an immune reaction. Most antigens are proteins or pieces of proteins from an invader, though sugars and other molecules can serve too. When you hear "antigen," think "the specific thing the immune system is reacting to."

Here's the core problem the adaptive system has to solve. The T cells you met earlier, the coordinators and the killers of adaptive immunity, can't recognize a free-floating invader on their own. A T cell only recognizes its target when that target is chopped into a small piece and displayed on the surface of one of your own cells, held up in the grip of an MHC molecule. This displaying of antigen fragments on MHC is called antigen presentation, and it's the event that kicks off the entire adaptive response. Without presentation, T cells stay blind, and adaptive immunity never launches.

Why would the immune system build in such a strange requirement? Two scientists, Rolf Zinkernagel and Peter Doherty, worked out the answer in the early 1970s with a set of clever experiments in mice infected with a virus. They discovered that a cytotoxic T cell could only kill an infected cell if that cell carried the same MHC type as the T cell itself. The T cell wasn't recognizing the virus alone. It was recognizing the virus and the self-MHC molecule together, as a package. This principle, called MHC restriction, explained how T cells keep their attention on your own cells and only strike when one of those cells is displaying something foreign. Zinkernagel and Doherty shared a Nobel Prize for the discovery in 1996. Their finding revealed that MHC molecules aren't just ID tags, as we described them in Chapter 1. They're active display cases that show T cells what's happening inside and around your cells.

Your body runs two separate presentation systems, and the difference between them is one of the most useful things to understand in all of immunology. They rely on two classes of MHC molecule, cleverly named MHC class I and MHC class II, and each class reports a different kind of news to a different kind of T cell.

MHC class I molecules answer the question "what am I making inside myself?" Nearly every cell in your body carries MHC class I, and each one constantly grabs small samples of the proteins being manufactured in its interior, brings them to the surface, and displays them. A healthy cell shows only normal, self proteins, and T cells passing by see nothing to worry about. But a cell infected by a virus is forced to manufacture viral proteins, and it can't help displaying fragments of them on its class I molecules along with everything else. This is the cell inadvertently confessing that it's been hijacked. Cytotoxic T cells, the CD8 killers, are the ones that read MHC class I. When they spot foreign fragments on a cell's class I display, they know that cell has been compromised, and they destroy it. Because every nucleated cell reports its own interior this way, class I presentation is a surveillance system that lets your body catch infections and cancers hiding inside otherwise normal-looking cells.

MHC class II molecules answer a different question: "what did I find in the neighborhood?" Only a small set of specialized cells carry MHC class II, the professional antigen-presenting cells you met in the last chapter, chiefly dendritic cells, macrophages, and B cells. These cells engulf invaders from their surroundings, break them into pieces, and load those pieces onto their class II molecules to display at the surface. This is a cell reporting on a threat it found outside itself, not inside. Helper T cells, the CD4 coordinators, are the ones that read MHC class II. When a helper T cell recognizes a foreign fragment on a class II display, it knows an invader is present in the area and swings into action to organize the response. Class II presentation, then, is how the immune system alerts its coordinators that there's a problem to manage.

There's a tidy memory trick for keeping the pairings straight, sometimes called the rule of eight. CD8 cytotoxic T cells go with MHC class I, and eight equals one times eight. CD4 helper T cells go with MHC class II, and eight equals two times four. Whenever you get tangled up about which T cell reads which MHC class, the rule of eight sorts it out.

Feature	MHC Class I	MHC Class II
Found on	Nearly all nucleated cells	Professional antigen-presenting cells only
Displays antigens from	Inside the cell (self and viral proteins)	Outside the cell (engulfed invaders)
Read by	Cytotoxic (CD8) T cells	Helper (CD4) T cells

Reports	"What I'm making inside"	"What I found outside"
Result of foreign display	Infected cell gets killed	Helper T cell coordinates a response

This split solves a hard logistical problem. Threats can be inside your cells, like viruses, or outside them, like most bacteria, and the two locations call for different responses. By running two presentation systems aimed at two different T cell types, your body can tell the difference between an internal infection and an external one, and respond to each appropriately. A virus hiding inside a lung cell gets reported through class I to the killers who can destroy that cell. A bacterium swimming in your tissue gets captured, reported through class II to the coordinators, and dealt with by the broader response they organize.

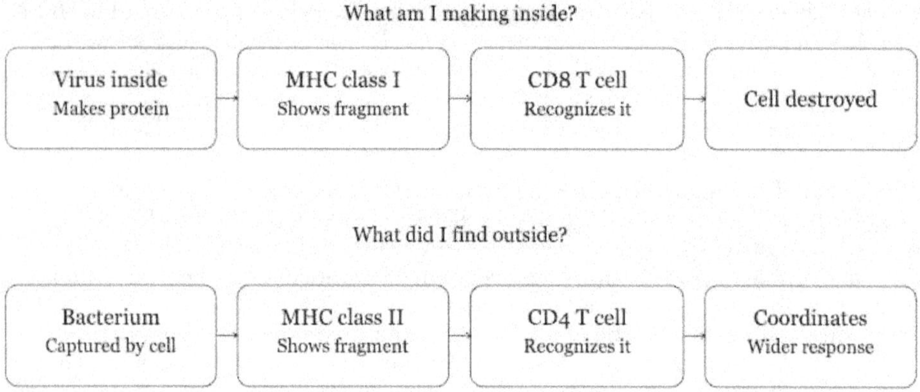

Location of the threat decides which lane runs

Dendritic cells have one more talent worth mentioning, because it fills what would otherwise be a dangerous gap. Normally, a cell can only alert killer T cells to viruses infecting that same cell, through class I. But what if a virus infects a cell that isn't good at activating T cells in the first place? Dendritic cells solve this with a maneuver called cross-presentation. They can take antigens they've engulfed from outside and, unusually, load them onto their own class I molecules to activate cytotoxic T cells directly. This lets dendritic cells kick-start the killer response against viruses even when they themselves weren't the ones infected, and it's one more reason these cells sit at the center of launching adaptive immunity.

T Cell Activation and Differentiation

Antigen presentation sets the stage, but presentation alone doesn't switch a T cell on. Activating a T cell is a deliberate, multi-step process with built-in safety checks, and for good reason. A T cell, once activated, can unleash a great deal of destruction, so your body doesn't turn one on casually. It demands more than one confirmation before committing.

The first thing to understand is how rare the right T cell is. Your body carries an enormous variety of T cells, each one bearing a receptor tuned to a different specific antigen, so that somewhere in the crowd there's a specialist for almost any invader. But for any given threat, only a tiny handful of your T cells happen to have the matching receptor. The odds of the right cell bumping into the right antigen by chance would be hopeless if it happened randomly throughout your body. This is exactly why the lymph nodes from Chapter 1 exist. Dendritic cells carry captured antigens to the lymph nodes, and T cells constantly circulate through those same nodes, so the nodes act as meeting points where the rare matching cell and its target are brought together efficiently. A T cell may pass through many lymph nodes, sampling the antigens on display, until it finds the one it was built to recognize.

When the matching T cell finally meets its antigen, activation requires two separate signals, and the two-signal requirement is one of the most important safety features in your immune system. The first signal is recognition itself: the T cell receptor binds the antigen presented on an MHC molecule. This confirms that the T cell has found its specific target. But signal one alone isn't enough. The T cell also demands a second signal, called costimulation, a separate confirming handshake between other proteins on the T cell and the antigen-presenting cell. Only when a T cell receives both signals at once does it fully activate.

Why insist on two signals instead of one? Because the second signal is a danger check, and it links the adaptive system back to the innate one in an elegant way. An antigen-presenting cell only displays the costimulatory second signal strongly when it has detected genuine danger, when its pattern recognition receptors from the last chapter have sensed the molecular signatures of a real invader. If a T cell recognizes an antigen, signal one, but there's no danger signal, no costimulation, the T cell interprets this as a false alarm. Rather than activating, it shuts down, becoming unresponsive to that antigen in a state called anergy, or it dies. This is a safeguard against autoimmunity. Your own harmless proteins are present all the time, and T cells encounter them constantly, but because those self proteins don't trigger danger signals, the T cells that recognize them get switched off rather than turned loose. The two-signal system helps ensure your T cells only attack when there's a real threat, not just because they happened to recognize something.

There's often a third signal too, and it determines what kind of cell the activated T cell becomes. This third signal comes from cytokines, the chemical messengers released by the antigen-presenting cell and the surrounding tissue. These cytokines carry information about the type of threat, and they steer the newly activated T cell down one specialized path or another. A viral infection produces a different cytokine mix than a parasitic worm, and that difference tells the T cell what kind of response to mount.

Once a T cell receives its signals and commits, it does something dramatic: it multiplies. This is called clonal expansion, and it's the answer to the rarity problem. That single matching T cell, or the few of them, divides rapidly again and again, at times as often as once every six hours, producing thousands of identical copies within a few days. All of these copies carry the same receptor, so they're all specialists against the same invader. In under a week, your body goes from a handful of the right T cells to an army of them, all aimed at the specific threat.

The idea behind this has a name and a history. In 1957, an Australian scientist named Frank Macfarlane Burnet proposed what he called clonal selection theory, and it reshaped how everyone thought about immunity. Burnet's insight was that your body makes a vast diversity of lymphocytes before ever encountering an invader, each with its own unique receptor, and that an antigen simply selects the cells that already happen to match it, prompting those specific cells to multiply. The immune system doesn't design a custom receptor in response to a threat. It generates enormous variety in advance and lets the threat pick out the pre-existing cells that fit. Burnet shared a Nobel Prize in 1960 for related work on immune tolerance, and clonal selection remains one of the foundational concepts of the field. Every time you fight off an infection, you're watching clonal selection and expansion play out.

As the activated T cells multiply, they also specialize, a process called differentiation, and here the two main T cell types diverge along different paths.

Cytotoxic T cells, the CD8 killers, differentiate into effector cells armed and ready to destroy infected targets. Once mature, they leave the lymph node, circulate through the body, and hunt for cells displaying their target antigen on MHC class I. When they find one, they kill it using perforin and granzymes, the same execution method you saw with natural killer cells in the last chapter. One activated cytotoxic T cell can kill many infected cells in sequence, moving from one target to the next, which makes them formidable against viral infections where many of your cells have been hijacked.

Helper T cells, the CD4 coordinators, are more complicated, because they don't all become the same thing. Guided by that third cytokine signal, a helper T cell differentiates into one of several distinct subtypes, each tailored to a particular class of threat. This is where the immune response gets customized to the enemy at hand.

The main helper subtypes are worth knowing by name, because they come up throughout immunology and medicine. Th1 cells specialize in fighting invaders that hide inside your cells, like viruses and certain bacteria. They pump out cytokines that activate macrophages and support the cytotoxic T cell killers, sharpening the response against internal threats. Th2 cells specialize in fighting large parasites, like worms, and they also drive the responses involved in allergies. They release cytokines that recruit specialized cells against parasites and direct B cells toward certain antibody types. Th17 cells specialize in fighting bacteria and fungi outside your cells, and they work largely by summoning neutrophils to the site. A fourth type, follicular helper T cells, specialize in helping B cells produce better antibodies, and we'll see them at work in the next section. And the regulatory T cells you met earlier act as the brakes, holding the whole response in check.

Helper T Subtype	**Specializes Against**	**Main Action**
Th1	Viruses and intracellular bacteria	Activates macrophages and cytotoxic T cells

Th2	Parasites like worms; drives allergy	Recruits anti-parasite cells; directs B cells
Th17	Extracellular bacteria and fungi	Summons neutrophils to the infection
Follicular helper (Tfh)	Assists B cells	Helps B cells make higher-quality antibodies
Regulatory (Treg)	Overactive responses	Suppresses and controls the immune reaction

This branching is why the immune system can respond so differently to different threats. A worm and a virus are nothing alike, and they call for completely different defenses. By reading the cytokine signals and differentiating into the matching subtype, helper T cells make sure your body brings the right kind of response to each specific kind of invader, rather than mounting the same generic attack against everything.

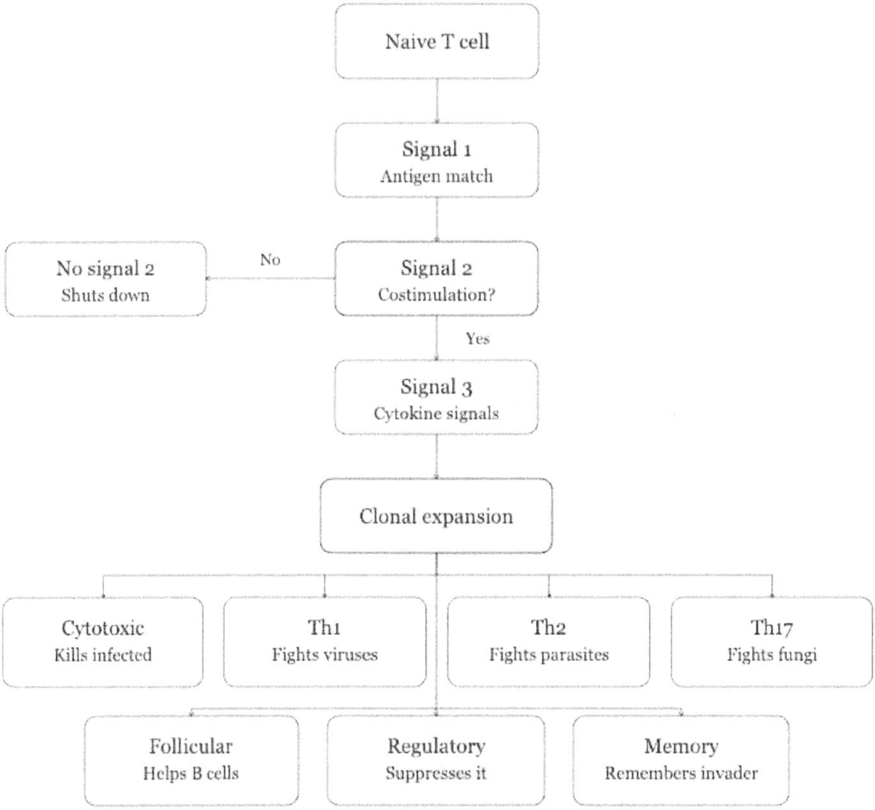

The right response for each specific invader

B Cell Activation and Antibody Classes

While T cells are being activated, the B cells are getting to work too, and their activation follows its own path toward a single goal: producing antibodies precisely tuned to the invader. You already know from Chapter 2 that an activated B cell becomes an antibody-producing plasma cell. Now we'll look at how a B cell gets switched on in the first place, and at the different types of antibodies it can make, because those types matter enormously.

Most B cell activation depends on help from T cells, and the partnership is a nice piece of teamwork. A B cell recognizes its specific antigen directly, latching onto a free-floating invader with its surface receptor. But recognition usually isn't enough to fully activate it. After binding the antigen, the B cell engulfs it, breaks it into pieces, and displays those pieces on its own MHC class II molecules, acting as an antigen-presenting cell in its own right. A follicular helper T cell, one that recognizes the same invader, then reads that display and provides the B cell with confirming signals and cytokines. Only with this T cell help does the B cell become fully activated. This arrangement, called a T-dependent response, means a B cell and a helper T cell recognizing the same invader have to agree before antibody production ramps up, another double-check against mistaken attacks.

Some antigens can activate B cells without any T cell help at all, a route called a T-independent response. Certain invaders, particularly bacteria with surfaces made of repeating sugar units, can trigger B cells directly because their repetitive structure engages many of the B cell's receptors at once. These responses are faster but weaker: they produce antibodies with little refinement and generate poor memory.

This has a real medical consequence. Young children under about two years old don't mount good T-independent responses, which is why plain sugar-based vaccines don't protect them well, and why vaccine designers had to develop cleverer versions to guard infants against certain bacteria. We'll return to that in the vaccination chapter.

When B cells do get full T cell help, something remarkable happens inside the lymph nodes. The activated B cells gather into specialized structures called germinal centers, and within these centers they undergo a process that improves their antibodies over time. The B cells begin mutating the genes for their antibodies at a furious rate, a process called somatic hypermutation. Most of these mutations do nothing useful or make the antibody worse, but a few, by chance, produce an antibody that grips the invader more tightly than before. The B cells with these improved, higher-affinity antibodies get selected to survive and multiply, while the ones with weaker antibodies die off. Round after round, this selection ratchets up the quality of the antibodies being made, a process called affinity maturation. By the end of a response, your body is producing antibodies that bind the invader far more tightly than the ones it started with. Your immune system, in effect, runs a rapid breeding program for better antibodies in the middle of every serious infection.

Germinal centers do one more thing that changes the whole character of the response: they let B cells switch which class of antibody they produce. Every antibody has the same basic job of recognizing a specific target, but antibodies come in several different classes, or isotypes, each built for a different setting and a different task. A B cell can change the class of antibody it makes while keeping the exact same target specificity, a process called class switching, and it's directed by cytokine signals from helper T cells depending on what kind of defense the situation calls for.

There are five antibody classes in humans, and knowing what each one does gives you a working understanding of how antibodies defend you across your whole body.

IgM is the first responder. When your body meets a new invader, IgM is the first antibody class it produces. IgM molecules join together in groups of five, forming a large, star-shaped cluster with many binding sites, which makes them good at grabbing invaders and clumping them together, and good at triggering the complement system from the last chapter. Because it appears early, the presence of IgM against a particular microbe is a sign of a recent or ongoing infection, which is something doctors use in diagnosis.

IgG is the workhorse. It's the most abundant antibody in your blood, making up the majority of the antibody in your circulation, and it's the main antibody of a mature, refined response. IgG does a bit of everything: it neutralizes invaders, tags them for phagocytes to eat, and activates complement. It also has a special ability that protects newborns. IgG can cross the placenta from mother to baby during pregnancy, so infants are born with a supply of their mother's IgG antibodies that guards them during their first months of life, before their own immune system is fully up to speed. This borrowed protection is one of the reasons a mother's own immunity matters so much for her newborn.

IgA is the guardian of your surfaces. It's the main antibody found in your secretions: your saliva, tears, sweat, mucus, and the lining of your gut and airways. IgA patrols the wet surfaces where invaders try to enter, neutralizing them before they can get through the barriers you read about in the last chapter. It's also present in breast milk, which means a nursing mother passes protective IgA to her baby, coating the infant's gut against infection. By quantity, your body actually produces more IgA than any other antibody class, precisely because it has so much surface area to defend.

IgE is the smallest player by amount but causes some of the most noticeable effects. Very little IgE circulates in your blood; instead, most of it sits bound to mast cells, the tripwire cells from the last chapter. IgE evolved mainly to fight large parasites like worms, triggering a forceful response against them. But IgE is also the antibody behind allergies. When IgE mistakenly targets a harmless substance like pollen or peanut protein, it triggers mast cells to release histamine, producing the sneezing, itching, and swelling of an allergic reaction, and in severe cases the dangerous whole-body reaction called anaphylaxis. We'll come back to allergies in a later chapter, but the culprit is IgE.

IgD is the least understood of the five. Small amounts appear on the surface of B cells that haven't yet been activated, where it seems to help those cells respond to antigen, but its full role is still being worked out. It's a reminder that even now, parts of the immune system remain incompletely mapped.

Antibody Class	Where It Works	Main Role
IgM	Blood, early in infection	First responder; clumps invaders; activates complement
IgG	Blood, throughout the body	Main all-purpose antibody; crosses the placenta
IgA	Secretions and mucosal surfaces	Guards entry points; passed in breast milk
IgE	Bound to mast cells	Fights parasites; drives allergic reactions
IgD	Surface of naive B cells	Helps B cells respond; role still unclear

Put affinity maturation and class switching together and you can see how much better the antibody response gets over the course of an infection. Your body doesn't just produce antibodies against the invader. It produces steadily better antibodies, in the right class for the job, refined and specialized as the response goes on. The antibodies you have at the end of a serious infection are sharper and better matched to the enemy than the ones you started with, and that improvement is exactly what your body remembers for next time.

Immunological Memory

Everything we've described so far, the presentation, the activation, the antibody refinement, produces a powerful response to an invader. But the response is temporary. Once the threat is defeated, the vast army of effector cells your body built is no longer needed, and most of them die off within a couple of weeks. Keeping millions of activated killer cells and antibody factories running forever would be wasteful and dangerous. So the response winds down, the cells clear out, and your body returns to a resting state.

Except it doesn't return all the way. This is the genius of the adaptive system, and it's what separates it from the innate defenses that forget everything. As the response resolves, a small fraction of the T cells and B cells that fought the infection don't die. Instead, they convert into long-lived memory cells and settle in for the long haul. These memory cells carry the exact receptors that recognized the invader, plus, in the case of B cells, the refined high-affinity antibodies produced through affinity maturation. They're the immune system's records office, holding onto the specific knowledge of a defeated enemy.

Memory comes in a few forms working together. Memory T cells persist for years, some patrolling your bloodstream and lymph nodes, others stationed permanently in the tissues where an invader is likely to return, like your skin, lungs, and gut, ready to respond right at the point of entry. Memory B cells persist too, standing by to spring back into antibody production the moment their target reappears. And there's a third piece that often gets overlooked: some plasma cells, the antibody factories themselves, become long-lived and take up residence in your bone marrow, where they quietly secrete antibodies for years or even decades. This is why, long after an infection or a vaccine, you can still have protective antibodies circulating in your blood without any active infection to produce them. A hidden population of plasma cells in your bones keeps the supply topped up.

How long does immune memory last?

It depends heavily on the invader. Immunity to some infections, like measles, can last essentially your whole life; people who had measles as children carry protective immunity many decades later. Immunity to others fades over years, which is why some vaccines require booster shots to refresh the memory, and why your protection against certain illnesses weakens with time. The durability of memory varies from one microbe to another, and immunologists are still working out exactly what makes memory against some threats last a lifetime while memory against others slips away. What's consistent is that memory, however long it lasts, is what turns a first, difficult encounter into a second, easy one.

Primary vs. Secondary Immune Responses

The whole point of immunological memory becomes clear when you compare what happens the first time your body meets an invader with what happens the second time. These two encounters are so different that immunologists give them separate names: the primary response and the secondary response. The gap between them is the reason vaccines work and the reason you rarely suffer through certain diseases twice.

The primary response is what happens on first contact, and it's slow. When a genuinely new invader breaches your defenses, your body has to find the rare matching lymphocytes, activate them, expand them into an army, and refine their antibodies, all the machinery we've walked through in this chapter. This takes time. For roughly the first five to seven days, there's a lag while all this gets underway, and during that lag you have little specific protection, which is why a new infection can make you feel sick for days before you recover. When antibodies finally do appear, IgM comes first, followed by IgG as the response matures. The antibody levels rise, peak at a modest level, help defeat the infection, and then decline over the following weeks as the response resolves. The primary response gets the job done, but it's neither fast nor especially strong, and you feel every bit of that delay.

The secondary response is what happens on any later encounter with the same invader, and it's a different experience entirely. Because your memory cells are already present, already specific to this exact threat, and already carrying refined antibodies, there's almost no lag. The memory cells recognize the returning invader within a day or two and reactivate immediately, skipping most of the slow buildup the primary response required. The result is faster, larger, and better in every measurable way. Antibody levels climb much sooner and reach a peak often ten to a hundred times higher than the primary response achieved. The antibodies are predominantly IgG rather than IgM from the start, and they bind the invader more tightly, thanks to the affinity maturation that happened the first time around. The invader is usually crushed before it can multiply enough to make you sick at all. You may fight off the entire infection without ever noticing it happened.

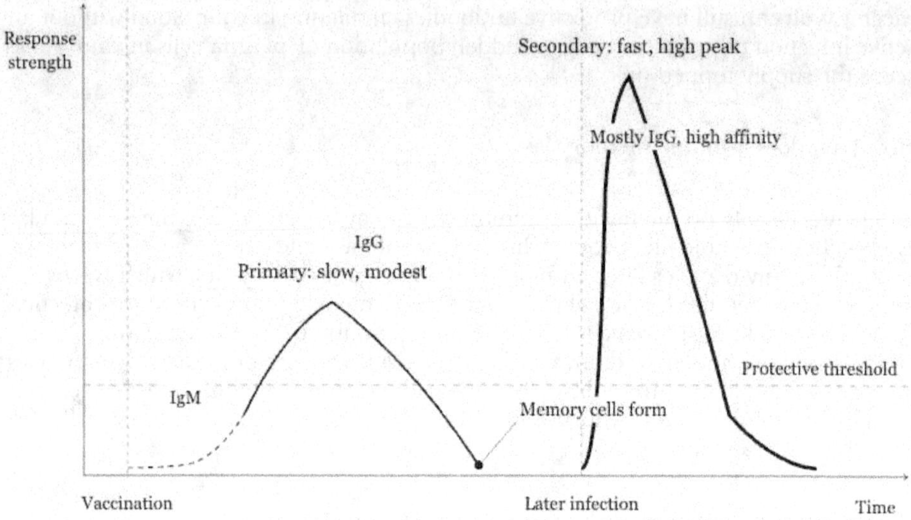

Feature	Primary Response	Secondary Response
When it occurs	First exposure to the invader	Later exposures to the same invader
Speed	Slow; 5 to 7 day lag	Fast; responds within 1 to 2 days
Antibody level reached	Modest peak	Much higher, often 10 to 100 times greater
Main antibody class	IgM first, then IgG	Mostly IgG from the start
Antibody quality	Lower affinity	Higher affinity, refined from before
How you feel	Often sick for days	Frequently no symptoms at all

This contrast is the entire foundation of vaccination, and seeing it laid out makes the logic obvious. A vaccine deliberately triggers a primary response using a harmless version or piece of an invader, one that can't make you sick. Your body goes through the slow buildup, forms memory cells, and then settles back down, exactly as it would after a real infection, but without the illness. Later, if you encounter the actual dangerous invader, your body treats it as a second exposure and mounts the fast, strong secondary response, defeating the threat before it can harm you. You get the protection of having already fought the disease without ever having to suffer through it. Booster shots work on the same principle: each additional exposure reinforces the memory and, for some vaccines, is needed to keep protection strong over time. The next chapter looks more closely at antibodies themselves, and a later chapter returns to vaccines in full, but the primary and secondary responses are the mechanism underneath all of it.

Discussion Question

The adaptive immune system builds an enormous diversity of lymphocytes in advance, before encountering any invader, so that a matching cell already exists for almost any threat it might one day face. This means your body spends resources maintaining vast numbers of cells that may never be used. Why might it make sense to prepare for threats you'll never meet, and what does this strategy reveal about the difference between how the innate and adaptive systems place their bets on an uncertain future?

Chapter 5: Antibodies and Antigens

You've now met antibodies several times. You know they're made by B cells, that they tag and neutralize invaders, and that they come in five classes. What you haven't seen yet is what an antibody actually is: its shape, how it grips its target, and why a single type of molecule can be aimed at nearly anything on Earth. This chapter opens up the antibody itself.

Antibodies are among the most studied and most useful proteins in all of biology. Your body makes billions of different ones, they're the basis of countless medical tests, and some of the best-selling drugs in the world are antibodies. To understand any of that, you have to start with how they're built, because the structure explains everything the antibody can do.

Antibody Structure

An antibody is a protein, and like all proteins it's a chain of amino acids folded into a specific shape. But antibodies have a shape so distinctive that it's become one of the icons of biology: the letter Y. That simple forked shape, with two arms branching off a single stem, is the key to how antibodies work, and once you understand the parts of the Y, the rest of antibody biology falls into place.

Every antibody is built from four separate protein chains, not one. There are two identical long chains, called heavy chains, and two identical short chains, called light chains. The two heavy chains form the backbone of the Y, running from the tips of the arms down through the stem. The two light chains sit alongside the upper part of the heavy chains, one on each arm. All four chains are held together by chemical links called disulfide bonds, strong connections between sulfur atoms that clamp the chains into the finished Y. Two heavy, two light, locked together: that's the basic architecture of every antibody in your body.

Now here's the feature that makes antibodies so remarkable. Each of these chains is divided into two kinds of region: a variable region and a constant region. The names tell you exactly what they mean. The constant region is essentially the same across all antibodies of a given class; it doesn't change from one antibody to the next. The variable region is different in every single antibody your body makes. This division is the heart of how antibodies achieve both diversity and function at once. The variable regions, sitting at the tips of the two arms, are where the antibody grabs its target, and their endless variation is what lets different antibodies recognize different invaders. The constant regions, making up the rest of the molecule and especially the stem, handle what happens after the target is caught.

Because the variable regions are so important, they've earned specific names. The very tip of each arm, where the variable region of a heavy chain meets the variable region of a light chain, forms the antigen-binding site. This is the business end of the antibody, the spot that physically latches onto the target. Since an antibody has two identical arms, it has two identical antigen-binding sites, one at the tip of each arm.

This means a single antibody can grab two copies of its target at once, which turns out to matter a great deal for clumping invaders together and holding them fast.

Zoom in on the binding site and there's even finer structure. Within each variable region are small stretches that vary far more wildly than the rest, called hypervariable regions or, more descriptively, complementarity-determining regions. There are three of these on the heavy chain's variable region and three on the light chain's, six in total, and together they form the actual surface that contacts the target. The word "complementarity" is the point: these regions fold into a shape that complements, or fits, the shape of the target, the way a mold fits the object cast in it. The other parts of the variable region, called framework regions, act as scaffolding that holds these six loops in the right position. When people say an antibody is specific to a particular invader, they're really talking about the precise shape of these six loops at the tips of its arms.

Scientists have names for the pieces of the Y too, and these come up constantly in immunology, so they're worth knowing. If you were to snip an antibody apart, the two arms are called Fab regions, short for "fragment, antigen-binding," because they contain the binding sites. The stem is called the Fc region, short for "fragment, crystallizable," a name that comes from a historical quirk of how the fragment behaved in the lab. The Fab arms do the recognizing. The Fc stem does the signaling. This division of labor is one of the most elegant things about the molecule: one end finds the enemy, the other end tells the rest of the immune system what to do about it.

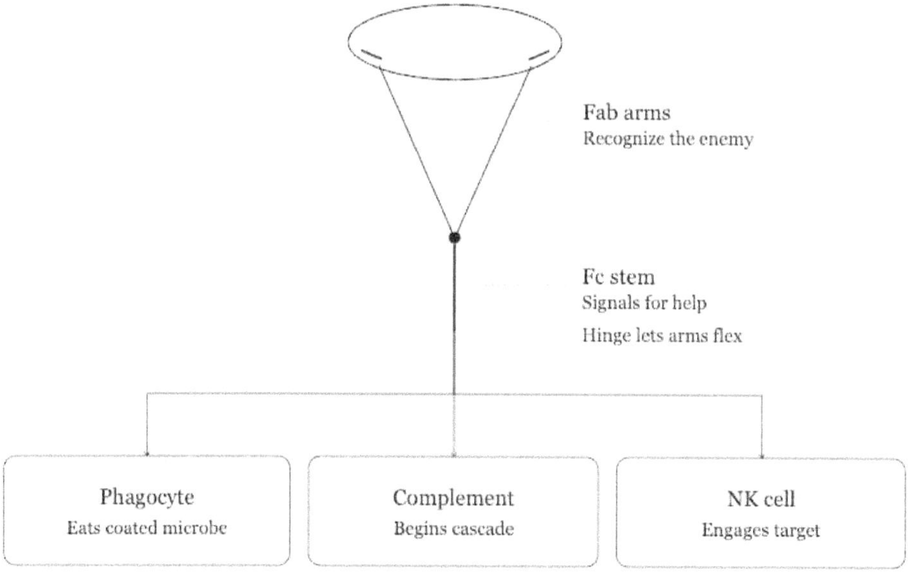

One end recognizes, the other connects to destruction

The Fc stem is what connects an antibody to the wider immune response, and it's the reason antibodies can do so much more than just stick to things. Immune cells like macrophages and neutrophils carry receptors that grab onto the Fc stem, so when an antibody coats an invader with its arms, its exposed stem becomes a handle that phagocytes can seize, making the invader easy to eat. The complement system from an earlier chapter also latches onto the Fc stem to begin its cascade. And the Fc region is what lets certain antibodies cross the placenta or get transported into secretions. So the constant stem, which doesn't vary and doesn't touch the target at all, is what determines an antibody's effect on the body. Change the stem, and you change what the antibody does, even if the arms are identical.

Between the arms and the stem, most antibodies have a flexible section called the hinge region. Just like a hinge on a door, it lets the two arms swing and flex relative to the stem. This flexibility isn't a minor detail. Targets on an invader's surface are spaced at varying distances, and the hinge lets the antibody's two arms spread wider or pull closer to grab two of them at once, whatever the spacing happens to be. An antibody with a stiff, fixed shape would be far less effective at latching onto both of its targets simultaneously. The hinge gives it the reach and adaptability to grip firmly.

The whole structure was worked out in the 1950s and 1960s by two scientists, Rodney Porter in Britain and Gerald Edelman in the United States, working separately. Porter used enzymes to carefully cut antibodies into fragments, discovering the Fab and Fc pieces and showing how they fit together. Edelman worked out the chain structure, demonstrating that the antibody was built from separate heavy and light chains linked by disulfide bonds. Between them, they revealed the four-chain Y that we now take for granted, and they shared a Nobel Prize for it in 1972. Before their work, nobody knew what an antibody looked like. After it, the Y-shaped model became the foundation for understanding the entire field, and eventually for designing antibodies as drugs.

Different antibody classes are built on variations of this same Y. The heavy chains come in different types, and the type of heavy chain an antibody uses determines its class and the exact structure it takes, which is a subject we'll return to shortly. But the core plan holds throughout: two heavy chains, two light chains, variable arms that recognize, a constant stem that signals, and a flexible hinge between them. Hold this picture in your head, because everything else about antibodies builds on it.

Antigen-Antibody Binding Specificity

The single most impressive thing an antibody does is recognize its target with extraordinary precision, ignoring the millions of other molecules around it. An antibody built to grab one particular protein will grab that protein and pass over nearly everything else, even molecules that look quite similar. This precision, called specificity, is the foundation of the entire adaptive immune system, and understanding how it works reveals one of the deepest tricks in biology.

Let's start with vocabulary, because two terms come up constantly and they name the two halves of the interaction. The specific part of an antigen that an antibody actually binds to is called the epitope. An antigen is often a large molecule, like a whole protein, but an antibody doesn't grab the entire thing. It grabs one small patch on the surface, and that patch is the epitope. On the antibody's side, the part that does the grabbing, that binding site at the tip of the arm, is called the paratope. So binding is a meeting of two specific surfaces: the paratope of the antibody fits the epitope of the antigen. A large invader like a bacterium has many different epitopes scattered across its surface, and your body makes many different antibodies, each recognizing a different one.

How does the binding actually happen? The classic description is a lock and key: the epitope is shaped so that only the correctly shaped paratope fits it, the way only the right key fits a lock. That image is roughly right and useful for a start, but the reality is a bit more fluid. Both the antibody and the antigen flex slightly as they come together, adjusting their shapes to fit more snugly, a phenomenon called induced fit. It's less like a rigid metal key and more like a handshake, where both hands mold slightly to each other. But the core idea holds: binding depends on the two surfaces being complementary in shape and chemistry, and if they don't match, they won't stick.

What holds the two surfaces together once they meet? Not a single strong bond, but many weak ones acting together, and this is a subtle point that explains a lot. The antibody and antigen are held together by several kinds of weak, non-covalent attractions: hydrogen bonds, electrical attractions between opposite charges, the faint pull between atoms in close contact, and the tendency of water-repelling regions to cluster away from water. Any one of these attractions is feeble and easily broken. But when the paratope and epitope fit closely, dozens of these weak bonds form all at once across the contact surface, and their combined strength holds the two molecules together firmly.

This is why shape matters so much – only when the fit is close can enough of these weak bonds form to add up to a strong grip. A poor fit leaves too few bonds, and the antigen slips away. Because these bonds are weak and non-covalent, the binding is also reversible, which the immune system needs; antibodies must be able to release and rebind as circumstances change.

Immunologists measure the strength of this binding with two related terms that are easy to confuse but genuinely different. Affinity refers to the strength of the bond between a single paratope and a single epitope, one arm gripping one target. Avidity refers to the total binding strength when an antibody grabs multiple targets at once with multiple arms. The difference matters enormously. An antibody with only modest affinity at each individual arm can still bind extremely tightly overall if it has many arms all gripping at the same time, because the total avidity is far greater than any single grip. This is exactly why the IgM antibody, which joins together into a cluster with ten arms, can hold onto an invader so effectively even early in an infection, before its individual arms have been refined for high affinity. Ten decent grips beat one good grip. The whole becomes far stronger than any of its parts.

Specificity isn't always perfect, and the exceptions matter for both disease and medicine. Sometimes an antibody built to recognize one epitope will also bind a different epitope that happens to have a similar shape, a phenomenon called cross-reactivity. Usually this is harmless or even helpful, letting an antibody recognize a slightly mutated version of an invader. But cross-reactivity can cause real trouble. If a microbe carries an epitope that resembles a molecule in your own body, the antibodies you make against the microbe might also attack your own tissue, a problem called molecular mimicry that contributes to certain autoimmune diseases. Rheumatic heart disease is a classic example: antibodies made against a streptococcal throat infection can cross-react with proteins in the heart, damaging the valves. The very specificity that makes antibodies so precise can misfire when two different things look too much alike.

Now we arrive at the deepest question of all, and it's a genuine puzzle. Your body can make antibodies against essentially any molecule, including synthetic chemicals that have never existed in nature and that your ancestors never encountered. How can a body possibly carry instructions for billions of different antibodies when your entire genome contains only around twenty thousand genes? There simply isn't enough DNA to store a separate gene for every possible antibody. For decades this seemed impossible, and it was one of the great unsolved problems in biology.

The answer, when it came, was startling, and it earned a Japanese scientist named Susumu Tonegawa a Nobel Prize in 1987. Tonegawa discovered that your body doesn't store complete antibody genes at all. Instead, it stores a modest collection of gene fragments, pieces that can be shuffled and combined in an enormous number of ways. When a B cell develops, it randomly selects one fragment from each of several groups and stitches them together to build a complete, unique antibody gene, a process called V(D)J recombination, named for the three groups of fragments involved. The number of possible combinations is staggering. Picture building sentences by picking one word from each of several long lists: a limited set of words in each list produces a nearly limitless number of sentences. By mixing and matching a manageable number of gene fragments, your B cells can generate more possible antibody variable regions than there are cells in your body.

The randomness goes even further. When the fragments are joined together, the cell adds and deletes small numbers of DNA letters at the junctions, introducing extra variation right at the point of the join, which happens to be exactly the spot that forms the center of the binding site. This junctional variation multiplies the diversity many times over. Then, because each antibody's binding site is formed by both a heavy chain and a light chain, and these are shuffled independently, the two randomly built chains combine to multiply the possibilities again. Stack all of this together and your body can produce something on the order of a trillion or more different antibody specificities, all from that small starting kit of gene fragments. This genetic shuffling, done before you ever meet an invader, is what fills your body with the vast diversity of B cells that clonal selection then draws upon.

There's a beautiful logic to how this connects to everything you've learned. The random shuffling creates enormous diversity in advance, ensuring that somewhere in your body there's a B cell whose antibody happens to fit almost any invader you might ever face. When an invader arrives, it selects the matching B cells, which multiply and refine their antibodies through the affinity maturation described earlier. So your antibody defense has two rounds of improvement: a random generation of diversity up front, giving you a rough-fit antibody for nearly anything, followed by a refinement process during infection that sharpens the fit. First breadth, then precision. That two-stage strategy is how a body with limited DNA defends against a practically unlimited variety of threats.

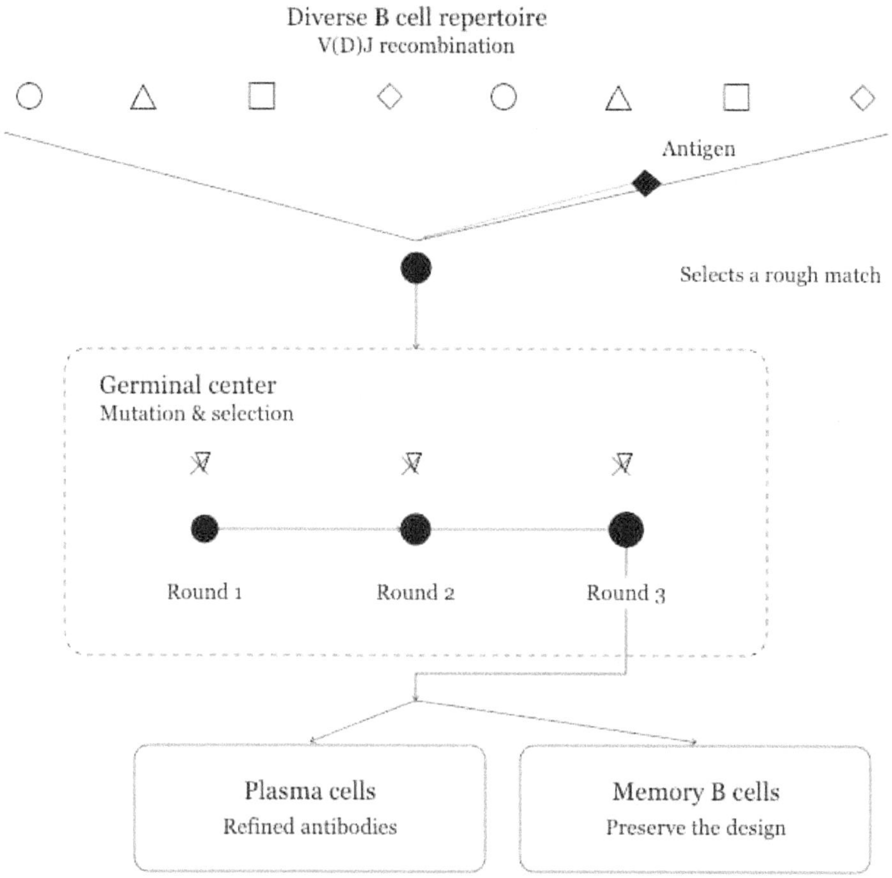

First breadth, then precision

Classes of Immunoglobulins

You already know the five antibody classes by what they do: IgM the first responder, IgG the all-purpose workhorse, IgA the guardian of surfaces, IgE the trigger of allergies, and IgD the mysterious one. Antibodies are also called immunoglobulins, which is where the "Ig" in each name comes from. What we haven't examined is why these classes differ structurally, and the answer comes straight back to the antibody architecture you just learned.

The class of an antibody is determined entirely by its heavy chain constant region, the stem of the Y. There are five types of heavy chain, and each one defines a class. The stem, remember, is the part that signals to the rest of the immune system, so changing the heavy chain changes what the antibody does after it binds, without changing what it recognizes. This is exactly why a B cell can switch classes while keeping the same target: it swaps out the heavy chain constant region for a different type while keeping the variable arms unchanged. Same recognition, different signaling stem, different effect.

The classes also differ in whether they travel alone or join together, and this shapes what each one is good at. IgG, IgE, and IgD each exist as a single Y-shaped unit, a monomer with two binding arms. IgM is different: five of these Y-shaped units link together into a large star-shaped cluster, a pentamer, held together by a small connecting protein called the J chain. This is what gives IgM its ten binding arms and its high avidity, the reason it grips invaders so well early on despite modest individual affinity. IgA usually travels as a single unit in your blood, but in your secretions, the saliva, tears, and gut fluid where it does most of its work, two units join into a pair called a dimer, again linked by a J chain.

IgA has one more structural feature that shows how well-suited it is to its job. When IgA gets transported out into your secretions, it picks up an extra piece called the secretory component, which wraps around the antibody and shields it from being digested. This matters because IgA works in harsh environments like the gut, full of enzymes that would tear an ordinary protein apart. The secretory component acts as armor, letting IgA survive and keep working in places where other antibodies would be destroyed. It's a small addition that makes surface defense possible.

Within some classes there are further subdivisions called subclasses, which fine-tune function even more. Human IgG comes in four subclasses, numbered IgG1 through IgG4, and they differ slightly in how strongly they activate complement and engage phagocytes. IgA comes in two subclasses. These distinctions get technical, but the practical point is that your body tailors its antibody response quite precisely, not just choosing a class but sometimes a specific subclass suited to the exact type of threat.

The classes differ dramatically in how long they last in your blood, which affects how they're used. Here's a comparison of the structural and practical features that distinguish them.

Class	Structure	Binding Arms	Approximate Blood Half-Life
IgG	Single unit (monomer)	2	About 3 weeks (longest)
IgM	Five units joined (pentamer)	10	About 5 days
IgA	Monomer in blood, dimer in secretions	2 or 4	About 6 days
IgE	Single unit (monomer)	2	About 2 days in blood
IgD	Single unit (monomer)	2	About 3 days

IgG's long three-week half-life is part of why it provides such durable protection and why it's the antibody that lingers after infections and vaccines. IgE's short blood half-life is misleading, though: while very little IgE circulates freely, the IgE bound to mast cells can persist for weeks, poised to trigger a reaction the moment its target appears. The structure and lifespan of each class fit the role it plays, which is a pattern you see throughout the immune system: form matched to function, down to the level of a single molecule.

Monoclonal vs. Polyclonal Antibodies

Here's a distinction that started as a technical point about how antibodies are made and turned into one of the biggest advances in modern medicine. The difference between polyclonal and monoclonal antibodies is the difference between a mixture and a pure product, and grasping it explains a whole category of drugs and diagnostic tests that shape healthcare today.

Start with what your own body does naturally. When you fight an infection, your immune response is polyclonal, meaning many different B cell clones activate at once, each recognizing a different epitope on the invader. A single bacterium has many epitopes across its surface, and your body makes a diverse collection of antibodies targeting all of them simultaneously. This diversity is a strength. By attacking the invader from many angles at once, a polyclonal response is hard for a microbe to escape; even if the invader mutates one epitope, antibodies against the others still work. The antibodies in your blood after any infection are always a polyclonal mixture, a whole crowd of different antibodies against the same enemy.

Medicine has used polyclonal antibodies for well over a century, and they still save lives. Antivenom for snakebites is a polyclonal product: animals, usually horses, are exposed to small, safe amounts of venom, they mount a polyclonal antibody response against it, and their antibody-rich serum is then purified and given to bite victims to neutralize the venom in their blood. The same approach protects against certain other toxins and infections. Another example is intravenous immunoglobulin, a preparation of pooled antibodies collected from thousands of blood donors, which contains a broad polyclonal mix and is used to treat patients whose own antibody production is deficient. During disease outbreaks, doctors have sometimes used convalescent plasma, the antibody-containing blood of recovered patients, on the same principle: borrowed polyclonal antibodies against a specific threat.

Polyclonal preparations have a drawback, though, and it's a serious one for consistency. Because they're a mixture harvested from living animals or donors, every batch is slightly different, and the exact composition can't be controlled or reproduced precisely. For research and for many treatments, scientists wanted something pure: a single, identical antibody, targeting a single epitope, that could be produced in unlimited quantity, exactly the same every time. That's a monoclonal antibody, one that comes from a single B cell clone and is therefore uniform. For a long time this was impossible, because a single B cell can't be grown indefinitely in the lab; it dies after a few divisions.

The breakthrough came in 1975, and it was ingenious. Two scientists, Georges Köhler and César Milstein, working in Cambridge, found a way to make an immortal antibody factory. They took a normal B cell, which makes the antibody you want but can't survive long, and fused it with a cancer cell, which divides forever but makes no useful antibody. The resulting hybrid cell, called a hybridoma, combined the best of both: it produced the exact antibody of the original B cell and it divided endlessly, churning out identical antibodies without limit. By selecting a hybridoma derived from one B cell, scientists could produce a limitless, perfectly uniform supply of a single antibody against a single chosen target. Köhler and Milstein shared a Nobel Prize for this in 1984, and their technique opened the door to an entire industry.

Monoclonal antibodies have become some of the most important drugs in the world, and you can spot them by their names. Most monoclonal antibody drugs end in the suffix "-mab," which stands for monoclonal antibody. Once you know to look for it, you'll notice these names everywhere in modern medicine. Rituximab targets a molecule on certain immune cells and treats some lymphomas and autoimmune diseases. Trastuzumab, sold as Herceptin, targets a protein found on some breast cancer cells and transformed the treatment of that disease. Adalimumab, sold as Humira, blocks an inflammatory signaling molecule and treats rheumatoid arthritis and other autoimmune conditions; for several years it was the best-selling drug on the planet. A newer group, including pembrolizumab and nivolumab, works against cancer in a different way, releasing a brake on the immune system so it can attack tumors, an approach we'll examine in the final chapter.

The technology kept improving in one particular way worth understanding. The earliest monoclonal antibodies were made in mice, which meant they were mouse proteins, and a patient's immune system would often recognize them as foreign and attack them, limiting how well they worked. So scientists learned to make monoclonals progressively more human. They built chimeric antibodies, part mouse and part human, then humanized ones with only a tiny mouse portion remaining, and finally fully human antibodies. This progression is encoded in the drug names too: the segment before "-mab" signals how human the antibody is. Making the antibodies more human reduced the chance the patient's body would reject them, which made them safer and more effective as long-term treatments.

Monoclonals also power a huge range of diagnostic tests, quietly, without most people realizing it. Home pregnancy tests use monoclonal antibodies to detect a specific hormone in urine. Rapid tests for infections, including the at-home tests that became familiar during recent disease outbreaks, use monoclonal antibodies to catch a specific viral protein and produce a visible line. Because each monoclonal recognizes exactly one target and behaves identically every time, it's ideal for a test that needs to give the same reliable answer in millions of uses. The precision that makes monoclonals good drugs also makes them good detectors.

Here's a summary of how the two types compare.

Feature	Polyclonal Antibodies	Monoclonal Antibodies
Source	Many B cell clones	A single B cell clone
Targets	Many epitopes on the antigen	One specific epitope only

Consistency	Varies batch to batch	Identical every time
Production	Harvested from animals or donors	Grown from immortal hybridoma cells
Examples	Antivenom, pooled immunoglobulin	Herceptin, Humira, cancer immunotherapies

The two aren't rivals so much as tools for different jobs. Polyclonal antibodies bring breadth and robustness, hitting a target many ways at once, which is what you want for neutralizing a complex venom. Monoclonal antibodies bring precision and reproducibility, hitting one exact target the same way every time, which is what you want for a drug or a diagnostic test. Both grew out of understanding the same basic molecule, the Y-shaped antibody, and learning to harness it deliberately.

Antigen Types and Sources

We've been using the word antigen throughout this chapter as "the thing an antibody recognizes," but it's worth pinning down more carefully now, because not everything the immune system can recognize provokes a response, and the differences have practical consequences. Sorting out what makes something an antigen, and where antigens come from, clarifies a lot about infection, allergy, transplantation, and vaccination.

First, a refinement of the term. An antigen is anything the immune system can recognize and bind to. But recognition and provocation aren't the same thing. A molecule that actually triggers an immune response, rather than merely being recognizable, is more precisely called an immunogen. All immunogens are antigens, but not every antigen can provoke a response on its own. This distinction sounds like hair-splitting, but it explains some real phenomena, as you'll see with small molecules in a moment.

What makes a molecule a good immunogen, one that reliably provokes a strong response? Several factors matter. Foreignness is the first: your immune system responds to things that differ from your own body, and the more foreign a molecule looks, the stronger the reaction it tends to draw. Size matters too: large molecules are generally far more immunogenic than small ones, because they offer more surface and more epitopes for the immune system to grab. Molecules below a certain size often provoke little or no response by themselves. Chemical complexity helps as well: a molecule with a varied, complicated structure gives the immune system more to recognize than a simple, repetitive one. Proteins tend to be the strongest immunogens because they're large and chemically complex, which is why most antigens the immune system cares about are proteins.

This brings up a fascinating category of small molecules called haptens. A hapten is a molecule too small to provoke an immune response on its own, but which can trigger one if it attaches to a larger carrier molecule, usually a protein. On its own, the hapten is invisible to the immune system. Bound to a big protein, it suddenly becomes part of something the immune system will react to, and antibodies form against the hapten. Karl Landsteiner, the same scientist who worked out blood types, did the foundational work on haptens in the early twentieth century.

Many drug allergies work exactly this way: penicillin is a small molecule that acts as a hapten, binding to proteins in your body and provoking an immune response that produces the allergic reaction. The itchy rash of poison ivy is another hapten reaction, caused by a small oily molecule from the plant that binds to proteins in your skin and triggers an immune attack. Small molecules that seem harmless become immunologically visible by hitching onto your own proteins.

Antigens are also classified by where they come from, and these categories map onto different kinds of immune responses. Exogenous antigens come from outside your body: the bacteria, viruses, pollen, and food proteins you encounter from the environment. These are the antigens your immune system engulfs and presents on MHC class II to alert helper T cells. Endogenous antigens are produced inside your own cells, typically when a virus hijacks a cell and forces it to make viral proteins, or when a cell turns cancerous and produces abnormal ones. These are the antigens displayed on MHC class I that alert the cytotoxic T cell killers. The location of the antigen, inside a cell or outside it, determines which arm of the response deals with it.

Two more categories fill out the picture and connect to earlier chapters. Autoantigens are your own molecules that mistakenly become targets of an immune response, which is what happens in autoimmune disease when the system loses tolerance for part of the body. Alloantigens are molecules that differ between individuals of the same species, which is why they matter for blood transfusions and organ transplants: the HLA proteins and blood group molecules that vary from person to person are alloantigens, and they're the reason a transplanted organ or mismatched blood gets attacked as foreign even though it comes from another human.

There's one unusual and dangerous class of antigen worth knowing about: superantigens. Normally, any given antigen activates only the tiny fraction of T cells specific to it. A superantigen breaks this rule. Certain bacterial toxins can bind T cells and antigen-presenting cells together in a way that bypasses the normal specific recognition, activating a huge proportion of your T cells all at once, regardless of what they're specific for. This triggers a massive, uncontrolled release of cytokines that can overwhelm the body. This is what happens in toxic shock syndrome, caused by toxins from certain staph and strep bacteria: the immune system, hijacked into overdrive by a superantigen, does far more harm than the infection itself. It's a reminder that an overwhelming immune response can be as deadly as the microbe that provoked it.

Finally, there's a way to make a weak antigen provoke a stronger response, and it matters enormously for the next section and for the vaccine chapter to come. A substance added to boost the immune response to an antigen is called an adjuvant. Adjuvants don't provide any target themselves; they act as a general alarm that heightens the immune system's reaction to the antigen they're mixed with. The most widely used adjuvant, aluminum salts, has been added to vaccines for decades to strengthen the response to the vaccine's antigen. By ramping up the danger signals that get the immune system's attention, adjuvants let a small amount of a purified antigen produce robust, lasting immunity.

Vaccines and Antigen Exposure

Everything in this chapter has been building toward a practical payoff, and here it is. Your immune system learns by exposure to antigens. It builds antibodies and memory against whatever antigens it encounters, and once you understand that, a powerful idea follows: if you could control which antigens the immune system meets, you could teach it to defend against a disease without ever suffering the disease. That's the entire concept behind vaccination, and it comes down to managing antigen exposure.

To see where vaccines fit, it helps to lay out all the ways a person can acquire immunity, because there are exactly four, and they divide along two simple lines. The first line is whether your own immune system does the work or whether you receive ready-made antibodies from elsewhere. Immunity your body builds itself, by mounting a full response and forming memory, is called active immunity. Immunity you get by receiving antibodies made by someone or something else is called passive immunity. The difference is enormous. Active immunity is slow to develop but long-lasting, because it comes with memory cells that persist for years. Passive immunity is immediate but temporary, because borrowed antibodies get used up and degrade within weeks to months, and no memory is left behind. You're renting protection rather than building it.

The second line is whether the immunity arises naturally, through ordinary life, or artificially, through a medical intervention. Cross these two lines and you get the four types of acquired immunity, which fit neatly into a grid.

	Active (your body builds it)	Passive (you receive antibodies)
Natural	Catching a disease and recovering	Antibodies from mother across placenta or in breast milk
Artificial	Getting a vaccine	Receiving an antibody injection, like antivenom

Walk through the four. Natural active immunity is what you get by catching an illness and recovering: your body meets the real invader, mounts a full response, and forms lasting memory, which is why you often don't catch the same illness twice. Natural passive immunity is the protection a baby receives from its mother, the IgG that crosses the placenta before birth and the IgA delivered in breast milk, both of which guard the infant temporarily while its own immune system matures. Artificial passive immunity is a medical version of the same borrowing: an injection of ready-made antibodies, like the antivenom or pooled immunoglobulin discussed earlier, or a monoclonal antibody treatment, all of which provide fast but short-lived protection with no memory. And artificial active immunity is vaccination: a deliberate, safe exposure to antigen that prompts your body to build its own lasting immunity, memory cells and all, without you having to get sick.

The genius of a vaccine is that it separates the antigen from the danger. Getting a disease gives you active immunity, but at the cost of suffering through the illness and risking its complications. A vaccine gives you the same active immunity by exposing your immune system to the antigen alone, or to a harmless version of the invader, so your body builds its memory response without ever facing the real threat. Your immune system can't tell the difference between practicing on a safe antigen and fighting the real thing; it forms memory either way. So after vaccination, you carry memory cells and refined antibodies against the invader, and if you ever meet the actual dangerous version, your body treats it as a repeat encounter and crushes it with the fast, strong secondary response described earlier. You gain the protection of having survived the disease without the disease itself.

This is also where adjuvants and antigen design come together. Because a vaccine often contains only a purified piece of an invader rather than the whole thing, that isolated antigen may not provoke a strong enough response on its own. Adding an adjuvant heightens the reaction, and choosing the right antigen, one that displays the epitopes your antibodies most need to recognize, shapes the quality of the immunity you develop. The whole enterprise rests on the antibody and antigen biology in this chapter: knowing what antigens are, how antibodies recognize their epitopes, and how exposure builds memory is what makes it possible to design a shot that trains your immune system on purpose.

How that antigen gets packaged into an actual vaccine, whether as a weakened live microbe, a killed one, a purified fragment, or a set of genetic instructions your own cells read to make the antigen themselves, is a subject with enough depth to deserve its own chapter, and we'll get there. What matters here is the underlying principle that ties this chapter to that one: immunity is learned through antigen exposure, and a vaccine is simply a way to arrange that exposure safely.

Discussion Question

Your body generates its enormous diversity of antibodies through random genetic shuffling, before it knows which invaders it will ever face. Randomness, by definition, also produces antibodies that could bind your own tissues. Given that risk, why might a system built on randomness be a better bet for defending against an unpredictable world than one that tried to build only the "right" antibodies from the start?

Chapter 6: The Immune System and Infection

Up to now, we've mostly talked about "invaders" and "microbes" as if they were all the same kind of enemy. They aren't. A virus, a bacterium, a fungus, and a parasitic worm are about as different from one another as a burglar, a swarm of ants, a patch of mold, and a tapeworm. They invade in different ways, hide in different places, and cause harm through different means. An immune system that fought all of them the same way would fail against most of them.

This chapter looks at infection from the enemy's side. For each major type of pathogen, we'll see how it attacks, how your body counters it, and the tricks it uses to survive. Then we'll close with fever, one of your body's oldest and most misunderstood weapons, which it turns against all of these invaders at once.

Viral Infection and Immune Evasion

Viruses are the strangest of all pathogens, and the hardest to fight, because of one defining fact: a virus can't reproduce on its own. It isn't even a complete living cell. A virus is essentially a set of genetic instructions wrapped in a protein coat, and it has no machinery of its own to copy itself. To reproduce, it has to break into one of your cells and hijack that cell's machinery, forcing your own equipment to build thousands of new viruses. This is what makes viruses so troublesome for your immune system. For much of their life cycle, they're hidden inside your own cells, shielded from antibodies and disguised as part of you.

Think about the problem this creates. Antibodies patrol your blood and tissues, but they can't reach inside a cell. Once a virus is in, floating antibodies can't touch it. So your body needs a two-part strategy: one defense to catch viruses while they're still outside cells, traveling between one victim and the next, and a completely different defense to deal with cells that have already been infected. Both parts run at once, and they cover for each other.

The first line against viruses is a chemical alarm you haven't met yet, and it's one of the most important antiviral tools your body has. When a cell detects that a virus has invaded it, it releases small signaling proteins called interferons, named because they interfere with viral reproduction. Two scientists, Alick Isaacs and Jean Lindenmann, discovered them in 1957. Here's the clever part. An infected cell, often before it dies, pumps out interferons that spread to neighboring cells and warn them that a virus is in the area. The neighboring cells respond by shifting into an antiviral state, ramping up defenses that make it much harder for the virus to take hold if it reaches them. In effect, an infected cell shouts a warning to everyone nearby, telling the whole neighborhood to lock its doors before the virus arrives. Interferons also rouse other immune cells, sharpening the wider response. This early warning buys time and slows the virus while the rest of the immune system gears up.

Next come the natural killer cells from an earlier chapter, which are especially useful against viruses in the first days of infection. They patrol for cells that look wrong and destroy infected cells on sight, without needing to be trained first, which makes them valuable during the window before the slower adaptive response is ready.

Then the adaptive system arrives with its two-part attack. Antibodies handle the viruses outside cells. When a virus is traveling through your blood or tissue between cells, antibodies bind to it and neutralize it, coating the virus so it can't attach to and enter a new cell. A neutralized virus is a stranded virus; it can't get inside, so it can't reproduce, and it eventually gets cleared away. This is exactly what most vaccines aim to produce: neutralizing antibodies ready to catch a virus before it can establish an infection. Meanwhile, the cytotoxic T cells handle the cells that are already infected. As you saw earlier, an infected cell displays fragments of viral protein on its MHC class I molecules, unwittingly confessing that it's been hijacked, and cytotoxic T cells recognize this and kill the cell, sacrificing it to stop the virus factory inside. Antibodies catch the virus in transit; killer T cells destroy the cells where it hides. Neither alone is enough, but together they can clear most viral infections.

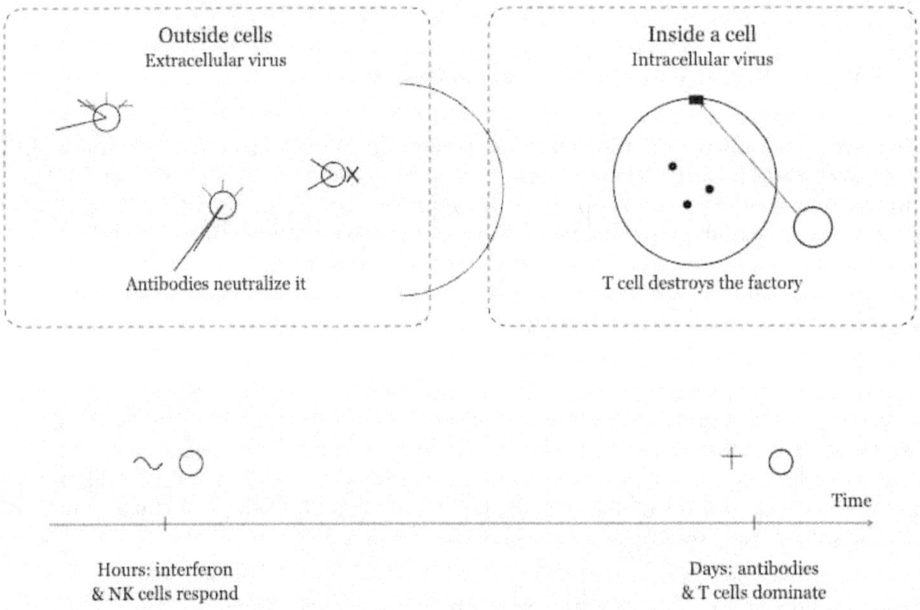

Viral location decides which weapon can reach it

So why do viruses still cause so much disease if we have all these defenses? Because viruses fight back, and they've evolved an impressive set of tricks to evade the immune system. Understanding these evasion strategies explains some of the most stubborn diseases we face.

The first trick is antigenic variation, or simply changing your appearance faster than the immune system can keep up. Antibodies recognize the specific shape of viral surface proteins, so if a virus changes those proteins, existing antibodies no longer fit. Influenza is the classic example. The flu virus accumulates small mutations in its surface proteins over time, a slow change called antigenic drift, which is why last year's antibodies, and last year's flu vaccine, don't fully protect you this year. The virus has quite literally changed its face. This is the reason a new flu vaccine has to be made every single year, updated to match the strains expected to circulate. Occasionally the flu virus undergoes a much larger, sudden change called antigenic shift, when different flu strains swap large chunks of genetic material and produce a virus so different that almost no one has immunity to it. That's how flu pandemics begin. HIV takes antigenic variation to an even greater extreme, mutating so rapidly that the virus population inside a single patient becomes a moving target, constantly shifting out from under the antibodies the body produces. This relentless variation is one of the main reasons an effective HIV vaccine has been so hard to make.

The second trick is latency, or hiding dormant where the immune system can't find you. Some viruses, after an initial infection, don't leave the body at all. Instead, they retreat into certain cells and go quiet, halting their reproduction and lying low for years. While a virus is latent, it makes few or no proteins, so there's nothing for the immune system to detect, and it becomes essentially invisible. The herpesvirus family are masters of this. The virus that causes cold sores hides in nerve cells between outbreaks, reactivating now and then, often when you're stressed or run down, to produce a fresh sore before going quiet again. The chickenpox virus does something similar and more dramatic: after you recover from chickenpox, the virus hides in your nerves for decades, and it can reawaken much later in life as shingles, a painful rash along the path of the affected nerve. The virus was there the whole time, waiting. Latency lets a virus outlast even a strong immune response, because you can't destroy what you can't find.

The third trick is directly sabotaging the immune defenses. Some viruses interfere with interferon signaling, muffling that early warning system so neighboring cells never get the alert. Others pull down the MHC class I molecules on the cells they infect, hiding the viral fragments that would otherwise flag the cell for cytotoxic T cells. This is a clever escape, but it runs into the counter-trap you read about earlier: a cell missing its MHC tags becomes a target for natural killer cells, which kill cells that have gone dark. The virus can hide from the T cells or hide from the NK cells, but hiding from both at once is nearly impossible, which is exactly why your body runs both systems.

The most devastating viral strategy of all is to attack the immune system itself, and this is what makes HIV so deadly. HIV specifically infects and destroys helper T cells, the CD4 coordinators that organize nearly the entire adaptive response. As the infection progresses untreated, the number of helper T cells steadily falls, and without these coordinators, the whole immune response unravels. B cells don't get the help they need, cytotoxic T cells aren't properly activated, and the system loses its ability to coordinate a defense. This is what defines AIDS: the immune system is so depleted that infections and cancers a healthy body would easily handle become life-threatening. The tragedy of HIV is that it turns the immune system's own command structure into its target. Modern antiviral drugs can't cure HIV, partly because it hides latently in some cells, but they can suppress it so effectively that a person's helper T cell count stays healthy and they can live a long, normal life. We'll touch on the immune consequences again when we discuss immunodeficiency.

Viral Evasion Strategy	How It Works	Example
Antigenic variation	Changes surface proteins so old antibodies no longer fit	Influenza, HIV
Latency	Goes dormant in cells, making few proteins to detect	Herpes, chickenpox/ shingles
Blocking interferon	Muffles the early antiviral warning to neighbors	Many viruses
Hiding MHC molecules	Pulls down the tags that flag infected cells to T cells	Some herpesviruses
Attacking immune cells	Infects and destroys the coordinators themselves	HIV

One more thing about viruses explains a familiar frustration. Why do you catch cold after cold, seemingly without ever becoming immune? Because the common cold isn't one virus but a large family of them, with well over a hundred different rhinovirus types alone, each with a different surface it presents to your antibodies. You do become immune to each one you catch, but there are so many, and new variants keep circulating, that you keep meeting ones you've never encountered. Your immune memory is working fine; it's just facing a nearly endless lineup of slightly different opponents.

Bacterial Infection and Host Defense

Bacteria are a completely different kind of enemy, and in one important way an easier one. Unlike viruses, bacteria are complete, living cells that can reproduce on their own, which means most of them multiply outside your cells, in your blood, tissues, and body fluids. And anything sitting outside your cells is something your antibodies and complement can reach directly. This changes the whole shape of the defense.

Against these extracellular bacteria, your body brings the tools you've already met, working as a coordinated team. Complement proteins tag the bacteria and can punch holes in some of them outright. Antibodies coat the bacteria, neutralizing them and marking them for destruction. And the phagocytes, the neutrophils and macrophages, do the actual eating, engulfing and digesting the marked bacteria.

The combination is powerful: complement and antibodies handle the marking, phagocytes handle the killing, and a bacterium coated in both becomes an easy, obvious meal. This is why the opsonization you read about earlier matters so much for bacterial defense. A naked bacterium is slippery and hard to catch, but a bacterium slathered in antibody and complement is grabbed and eaten in seconds.

Bacteria, of course, have their own defenses, and their evasion tricks differ from the viral ones. The most common is a physical shield. Many dangerous bacteria wrap themselves in a slippery outer capsule made of sugar molecules, which makes them hard for phagocytes to grab hold of. It's like trying to catch a greased ball. The bacterium that causes many cases of pneumonia and meningitis relies heavily on its capsule to resist being eaten, and the capsule is a major reason these bacteria can be so dangerous. Your body's counter to the capsule is antibodies: antibodies that bind the capsule give phagocytes something to grip, which is precisely why vaccines against these bacteria target the capsule and why they've prevented enormous numbers of childhood deaths.

The second bacterial weapon is chemical warfare, in the form of toxins. Many of the most harmful bacteria cause disease not by the infection itself but by releasing poisons. Some bacteria secrete toxins called exotoxins that damage your cells or disrupt your body's functions from a distance. Tetanus is caused by a toxin that locks your muscles into painful spasms. Botulism is caused by a toxin so potent that a tiny amount can be fatal, paralyzing muscles by blocking nerve signals. Diphtheria and cholera also do their damage through secreted toxins. Your body defends against these with a special class of antibodies called antitoxins, which bind the toxin and neutralize it before it can act. Several of our oldest and most successful vaccines, including the tetanus and diphtheria shots, work by training your body to produce these antitoxin antibodies. A different kind of bacterial poison, called endotoxin, isn't secreted but is part of the bacterial cell wall itself, the LPS molecule from an earlier chapter, and it's released when the bacterium dies. In small amounts it usefully alerts the immune system, but when large numbers of these bacteria are present, the flood of endotoxin can trigger a dangerous, body-wide overreaction, which brings us to one of the deadliest things the immune system does.

That overreaction is sepsis, and it's a sobering reminder that your defenses can harm you as much as any microbe. When a bacterial infection spreads into the bloodstream and the immune system detects danger everywhere at once, it can launch a massive, uncontrolled release of inflammatory cytokines throughout the entire body, sometimes called a cytokine storm. Instead of a focused response at one infected site, the whole body becomes inflamed. Blood vessels everywhere leak and dilate, blood pressure collapses, and organs begin to fail from lack of proper blood flow. This is septic shock, and it can kill within hours. The cruel irony is that in severe sepsis, the immune response itself, not just the bacteria, is doing the damage. Sepsis kills millions of people worldwide every year and remains one of the leading causes of death in hospitals. It shows, in the starkest terms, why controlling and shutting down an immune response is as important as starting one.

Not all bacteria play by the extracellular rules, and the ones that don't force your body to switch strategies entirely. Some bacteria have learned to survive inside your cells, often inside the very macrophages sent to destroy them. The bacterium that causes tuberculosis is the most important example. When a macrophage engulfs the TB bacterium, the bacterium blocks the macrophage from digesting it and instead sets up residence inside, turning its would-be killer into a hiding place. Antibodies and complement are useless against a bacterium hiding inside a cell, just as they're useless against a virus there. So the body falls back on cell-mediated immunity: helper T cells of the Th1 type release signals that supercharge the infected macrophages, boosting their killing power enough to destroy the bacteria they harbor. When even this can't clear the infection, the body walls the bacteria off instead, surrounding them with a dense ball of immune cells called a granuloma that contains the infection without eliminating it. This is why tuberculosis so often becomes a chronic, latent infection: a substantial share of the world's population, by some estimates around a quarter, carries the TB bacterium in a dormant, walled-off state, kept in check by the immune system but never fully destroyed. If that person's immunity later weakens, the bacteria can break out and cause active disease, which is one reason TB and HIV are such a dangerous combination.

There's a growing problem with bacterial infection that has nothing to do with your immune system and everything to do with our medicines. For decades, antibiotics have given us a reliable way to help the body clear bacterial infections. But bacteria evolve quickly, and every use of an antibiotic selects for the rare bacteria that happen to survive it, which then multiply and spread their resistance. The result is a rising tide of antibiotic-resistant bacteria, including strains like MRSA that shrug off drugs which used to kill them easily. This matters immunologically because when antibiotics fail, your immune system is left to fight these infections largely on its own, and for someone whose immunity is already compromised, that can be a losing battle. Antibiotic resistance is one of the more serious public health threats of our era, and it's a reminder that our medical tools work alongside the immune system rather than replacing it.

Fungal and Parasitic Immunity

Fungi and parasites round out the cast of pathogens, and while they cause less everyday illness than viruses and bacteria in wealthy countries, they're responsible for an enormous burden of disease worldwide, and they reveal parts of the immune system the other pathogens don't.

Start with fungi. Fungi are everywhere in the environment, and you breathe in fungal spores constantly, yet you rarely get a fungal infection. That's because a healthy immune system handles fungi with ease, mostly through neutrophils and the Th17 helper T cells that summon them. The telling thing about fungal infections is when they happen: they strike hardest in people whose immune defenses are weakened. This is why serious fungal infections are so common in patients with advanced HIV, cancer patients whose neutrophil counts have been knocked down by chemotherapy, and transplant recipients on immune-suppressing drugs. Fungi that a healthy body brushes off can become life-threatening when the immune system is down, which is why they're often called opportunistic infections, waiting for an opening. The yeast that causes common conditions like oral thrush and vaginal yeast infections lives on many people harmlessly but overgrows when immunity or the normal balance of microbes is disturbed. A mold that's harmless to most people can invade the lungs of someone with few neutrophils. And a particular fungal pneumonia became one of the defining illnesses of the AIDS epidemic, striking patients precisely because their helper T cells were gone. The clearest lesson from fungal infection is how much your everyday health depends on immune defenses you never notice working.

Parasites are a broad category, and they split into two very different groups that the immune system handles in different ways. The first group is protozoa, single-celled parasites like the ones that cause malaria and a range of other tropical diseases. Being single cells, they can sometimes be handled by mechanisms similar to those used against bacteria and viruses, though they're notoriously good at evasion. The second group is the helminths, the parasitic worms, and these are something else entirely: large, multicellular animals that can be millimeters or even meters long, living inside your gut, blood, or tissues.

Worms pose a problem your immune system's usual tools can't solve, and the solution is genuinely different from anything covered so far. A phagocyte can't eat a worm; the worm is thousands of times too big to engulf. Antibodies and complement can coat a worm's surface, but coating won't kill something that large. So the body deploys a specialized defense built around a cell you've only heard mentioned in passing: the eosinophil. Directed by the Th2 helper response and the IgE antibody class, eosinophils gather around a parasite that's too big to swallow and release their toxic internal contents directly onto its surface, essentially spraying it with poison from the outside to damage and kill it. IgE antibodies coat the worm and give the eosinophils something to bind, focusing the attack. This whole system, the Th2 response with its IgE, mast cells, and eosinophils, evolved largely to fight large parasites, and it works by attacking from the outside what can't be eaten from within. It's the same machinery that, when it misfires against harmless substances, produces allergies, which is a connection we'll return to in the next chapter.

Malaria deserves special attention, because it's one of the deadliest infections in human history and a showcase of parasite evasion. Malaria is caused by a protozoan parasite spread through mosquito bites, and it kills hundreds of thousands of people every year, most of them young children in Africa. What makes it so hard to defeat is its slippery life cycle and its talent for hiding. After a mosquito injects the parasite, it first hides inside liver cells, out of reach of most of the immune response. Then it invades red blood cells, where it's largely shielded from immune attack, and multiplies inside them before bursting out to infect more. Worse, the parasite constantly changes the proteins it displays on the surface of infected red cells, an antigenic variation similar to what some viruses do, so that just as the immune system learns to recognize it, it changes its appearance. This combination of hiding inside cells and constantly shifting its surface is why people in malaria regions can be infected repeatedly and why a fully protective vaccine has been so difficult to develop, though the first malaria vaccines have recently begun to reach children after decades of effort.

Parasites in general are the great escape artists of the microbial world, and the reason is time. Many parasites have co-existed with human hosts for a very long time, and this long shared history has let them evolve elaborate ways to dampen, dodge, and manipulate the immune response. Some coat themselves in the host's own molecules to appear as self. Some actively suppress the immune reaction around them. Many establish chronic infections that last for years, having reached a kind of stalemate in which the immune system can't clear them but manages to keep them somewhat in check. This is quite different from the pattern with most viruses and bacteria, and it reflects how thoroughly parasites have adapted to living inside us.

Here's an overview of how your immune system's main tools line up against each type of pathogen. This table ties together much of what you've learned across several chapters.

Pathogen Type	Where It Lives	Main Defenses Against It
Viruses	Inside your cells	Interferons, NK cells, cytotoxic T cells, neutralizing antibodies
Extracellular bacteria	Outside cells, in tissue and blood	Complement, antibodies, phagocytes, antitoxins
Intracellular bacteria	Inside cells, often macrophages	Th1 cells activating macrophages; granulomas
Fungi	Surfaces and tissues; opportunistic	Neutrophils and Th17 cells
Worms (helminths)	Gut, blood, tissues; too big to eat	Eosinophils, IgE, mast cells, Th2 response

Look across that table and you can see why your immune system has so many different cells and mechanisms. Each type of enemy demands a different tool. The variety in your defenses matches the variety of the threats, and a body missing any one of these capabilities would be left open to the specific pathogens that capability handles.

Fever as an Immune Strategy

Almost everyone has spent a miserable night burning up with a fever, wishing it would break. It's natural to think of fever as part of the illness, a symptom to be stamped out. But fever isn't something the infection does to you. It's something your body does on purpose, a deliberate defensive move that's so ancient and so widespread across the animal kingdom that it must be genuinely useful.

Your body regulates its temperature with remarkable precision, and the control center is a region of the brain called the hypothalamus, which acts as a thermostat set to keep you around your normal temperature. Fever happens when your immune system deliberately raises the setting on that thermostat. When your body detects an infection, certain molecules called pyrogens, meaning fever-producers, signal the hypothalamus to reset its target temperature higher. Some pyrogens come from the microbes themselves, but the main ones are your own cytokines, the same inflammatory signals released during any immune response, which travel to the brain and trigger the temperature increase. The hypothalamus responds by resetting your thermostat upward, and your body then works to reach the new, higher target.

This explains the miserable pattern of chills that comes with a rising fever, which makes more sense once you understand the thermostat. When the set point jumps up, your actual body temperature is suddenly below the new target, so your brain thinks you're too cold, even though you may already be warm. That's why you shiver and feel freezing at the start of a fever, piling on blankets while your temperature climbs. Shivering generates heat to raise your temperature toward the new set point. Later, when the infection is winding down and the set point drops back to normal, the reverse happens: now your temperature is above target, your brain thinks you're too hot, and you sweat and feel flushed as your body sheds heat to cool back down. The sweating that signals a fever "breaking" is simply your thermostat returning to normal.

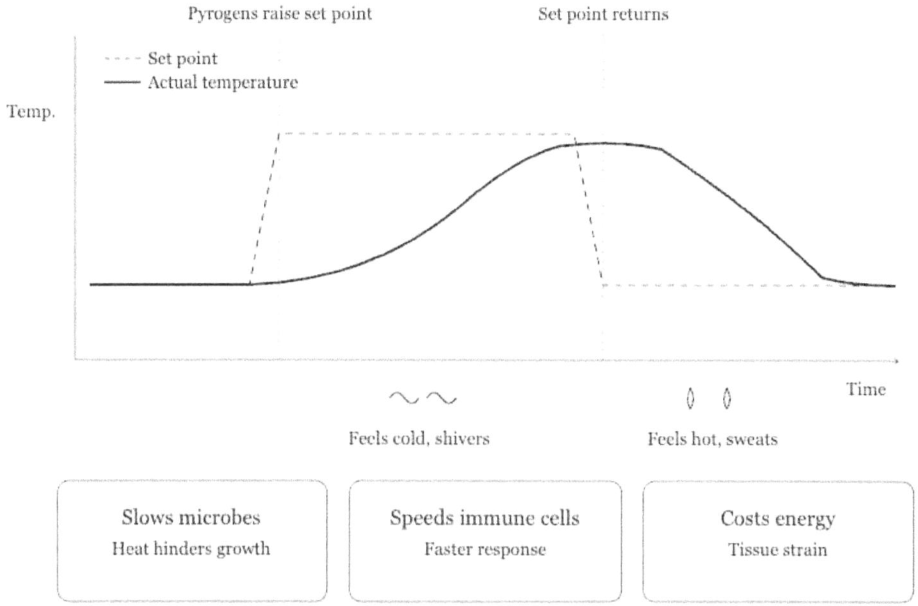

The gap between set point and actual temperature is what you feel

Why would raising your temperature help fight infection? Several reasons. Many bacteria and viruses reproduce best at your normal body temperature and grow more slowly when things heat up, so a fever creates a less hospitable environment for the invaders. At the same time, higher temperatures speed up your immune cells, making them move faster and work harder, so your defenses become more active just as the pathogens are slowed. Fever seems to tilt the battlefield in your favor on both sides at once, hindering the enemy while energizing the defense.

The best evidence that fever is truly protective comes from an unexpected place: cold-blooded animals. Reptiles can't generate their own body heat, so they can't produce a fever internally. But when researchers infected desert lizards with bacteria, the sick lizards deliberately sought out warmer spots to raise their body temperature, a behavior called behavioral fever, and the lizards that were allowed to warm themselves survived at much higher rates than those kept cool. An animal that can't make a fever will go out of its way to create one behaviorally, and doing so helps it survive. This kind of finding, repeated across many species, tells us fever has been conserved through hundreds of millions of years of evolution because it works.

None of this means fever is harmless or that every fever should be endured. Fever costs energy and makes you feel awful, which is part of why you rest when you're sick, and very high fevers can become dangerous, particularly for young children, who can suffer seizures, and for anyone when temperatures climb to extreme levels. This is why we have medicines to bring fevers down. Drugs like aspirin and ibuprofen reduce fever by blocking the production of prostaglandins, the molecules that carry the fever signal to the hypothalamus, which resets the thermostat back toward normal. Understanding that these drugs work by interrupting the fever signal, rather than by fighting the infection, clarifies what they do and don't accomplish. There's an ongoing discussion among doctors about whether it's wise to suppress a mild fever at all, given that fever appears to help fight infection, or whether the comfort of lowering it outweighs the modest benefit of letting it run. For a mild fever in an otherwise healthy person, the fever may be doing useful work, even as it makes for an unpleasant night.

Body's State	What the Thermostat Is Doing	What You Feel
Fever rising	Set point raised above current temperature	Chills and shivering; you feel cold
Fever steady	Body held at the new higher set point	Hot, achy, and tired
Fever breaking	Set point returns to normal, below current temperature	Sweating and flushing; you feel hot

Fever is a fitting note to end this chapter on, because it's a defense your body turns against every kind of pathogen we've discussed, viral, bacterial, fungal, and parasitic alike. It's not precise the way an antibody is precise. It's a broad, whole-body adjustment that makes conditions worse for invaders and better for your defenders, all at once. And like inflammation, it's a reminder that many of the things we experience as illness are actually the machinery of recovery at work.

Discussion Question

Fever, inflammation, and the aggressive immune responses to infection all make you feel sick, yet they're your body's own doing rather than direct damage from the pathogen. Given that feeling terrible often means your defenses are working hard, how should that change the way we think about treating symptoms, and where would you draw the line between a response worth enduring and one worth suppressing?

Chapter 7: Immune Dysfunction and Disease

Everything you've learned so far describes an immune system working the way it should. But the same system that protects you can turn against you, and when it does, the results range from annoying to fatal. An immune system can attack the wrong target, overreact to something harmless, fail to respond at all, or refuse to shut off. Each of these failures causes its own category of disease, and together they account for an enormous share of human suffering.

This chapter is about what happens when immunity goes wrong. We'll look at five kinds of dysfunction: the immune system attacking your own body, overreacting to harmless substances, failing from weakness, smoldering without resolution, and missing the cancer cells it's meant to catch. Each one is really the flip side of a defense you've already studied.

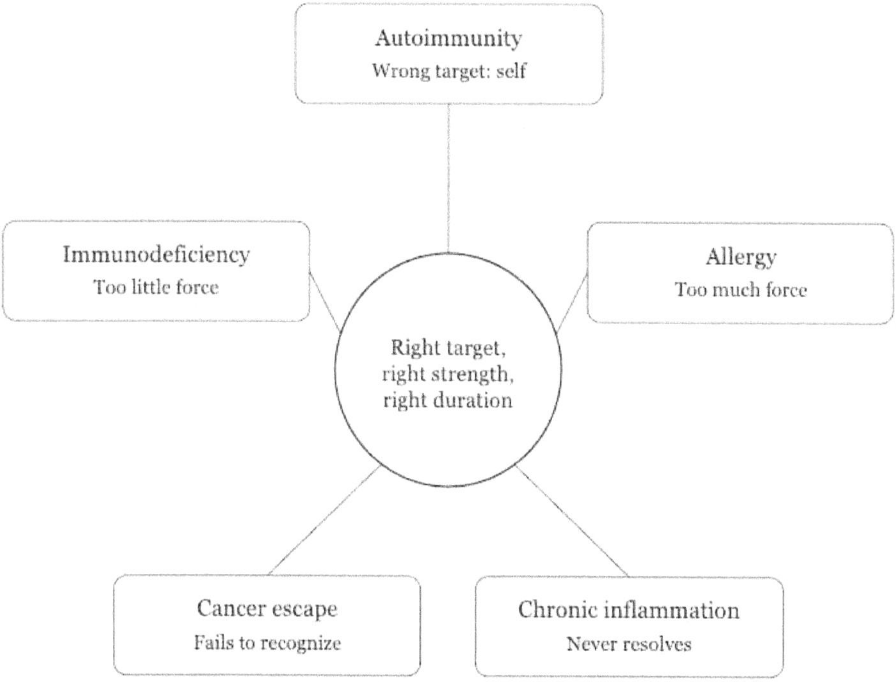

Five diseases, one problem: getting the calibration wrong

Autoimmune Disorders

The immune system's greatest challenge, as you learned in the very first chapter, is telling self from non-self. Autoimmune disease is what happens when it gets this wrong in the most dangerous direction: the system mistakes part of your own body for a foreign invader and attacks it. Your defenses, built to protect you, turn their full destructive power on your own tissues. The result depends entirely on which tissue gets attacked, which is why autoimmune diseases are so varied and often so puzzling.

To understand how this happens, recall the concept of tolerance, the immune system's trained ability to leave your own body alone. Your body works hard to build this tolerance. During their development, T cells that would react against your own tissues are screened out in the thymus, and B cells go through a similar weeding process, so that most self-reactive cells are eliminated before they ever enter circulation. This is called central tolerance, the removal of dangerous cells at their source. But the process isn't perfect, and some self-reactive cells inevitably slip through. So the body maintains a second line of defense called peripheral tolerance: mechanisms out in the body that keep any self-reactive cells that escaped from causing trouble, including the regulatory T cells you met earlier, which actively suppress cells that might attack the body. Autoimmune disease breaks out when both of these safeguards fail and self-reactive cells become active.

Why do these safeguards fail? The honest answer is that it's usually a combination of factors, and often we can't pinpoint exactly what tipped the balance in a given person. Genetics loads the dice: certain gene variants, especially particular HLA types, make some people far more prone to autoimmunity, which is why these diseases often run in families. But genes alone rarely cause autoimmune disease; something in the environment usually has to trigger it in a susceptible person. Infections are a common trigger, and one mechanism is the molecular mimicry you read about earlier, where a microbe carries a molecule resembling one of your own, so the antibodies and T cells you make against the infection also attack your tissue. Rheumatic heart disease, where a strep throat leads to an immune attack on the heart valves, is the classic case. Other triggers can include certain drugs, and possibly stress and other factors we don't fully understand.

One pattern stands out clearly across autoimmune diseases: they strike women far more often than men. The large majority of autoimmune disease patients are women, and for some conditions, like lupus, women outnumber men by roughly nine to one. The reasons aren't fully settled, but they likely involve the effects of sex hormones on the immune system and the fact that women carry two X chromosomes, which hold many immune-related genes. Whatever the cause, if you're trying to understand who develops these diseases, sex is one of the most reliable predictors.

Autoimmune diseases split into two broad groups based on how widely they strike, and this division helps make sense of their wildly different symptoms. Organ-specific autoimmune diseases attack one particular tissue, so the damage and symptoms are concentrated in that organ. Systemic autoimmune diseases attack targets found throughout the body, so they cause widespread, scattered symptoms affecting many organs at once. Knowing which group a disease falls into tells you a lot about what it does.

Consider some of the organ-specific diseases, because they show how the target determines the disease. In type 1 diabetes, the immune system destroys the insulin-producing cells in the pancreas, so the body loses its ability to control blood sugar and the person depends on insulin injections for life. It's important to understand that this is a completely different disease from the far more common type 2 diabetes, which isn't autoimmune at all; type 1 is an immune attack on the pancreas, and it often begins in childhood.

In multiple sclerosis, the immune system attacks the protective coating around nerve fibers in the brain and spinal cord, disrupting the signals that travel along them and causing problems with movement, sensation, and vision. The thyroid gland is a frequent target, and interestingly, it can go wrong in two opposite ways: in Hashimoto's thyroiditis, the immune system gradually destroys the thyroid, so it produces too little hormone and the person's metabolism slows down, while in Graves' disease, antibodies stimulate the thyroid to overproduce, speeding metabolism up. Same organ, opposite results, depending on whether the immune attack destroys or overstimulates.

Now consider the systemic diseases, which are harder to diagnose precisely because their effects are so scattered. Rheumatoid arthritis attacks the lining of the joints throughout the body, causing painful, swollen joints, and because it's driven by a body-wide immune process, it can affect other organs too. The most striking systemic disease is lupus, in which the immune system produces antibodies against components found in the nucleus of nearly every cell, so the disease can strike almost anywhere: the skin, joints, kidneys, heart, brain, and blood. Lupus is notoriously difficult to diagnose because its symptoms are so variable and can mimic many other conditions, and it often comes and goes in flares. In lupus, antibodies bind to self molecules and form clumps called immune complexes that deposit in tissues like the kidneys and trigger damaging inflammation, a mechanism we'll see again in the next section.

Disease	Type	Target of Attack	Main Effect
Type 1 diabetes	Organ-specific	Insulin-producing pancreas cells	Loss of blood sugar control
Multiple sclerosis	Organ-specific	Nerve fiber coating (myelin)	Disrupted nerve signals
Hashimoto's thyroiditis	Organ-specific	Thyroid gland	Underactive thyroid
Graves' disease	Organ-specific	Thyroid receptors	Overactive thyroid
Rheumatoid arthritis	Systemic	Joint linings, other tissues	Painful, swollen joints

Lupus	Systemic	Cell nucleus components, many organs	Widespread, variable damage

Treating autoimmune disease is genuinely difficult, and understanding why reveals a hard trade-off at the heart of immunology. The problem is that the immune system is attacking a target it shouldn't, but that same immune system is still needed to defend against real infections. You can't simply switch it off without leaving the person vulnerable. So most treatment relies on immunosuppression, drugs that dampen the immune response broadly, which reduces the autoimmune attack but also weakens protection against infection, leaving patients more prone to getting sick. Older treatments suppress immunity with a wide brush.

Newer treatments are more targeted: several of the monoclonal antibody drugs from the last chapter were developed to block specific inflammatory signals or deplete specific immune cells involved in a particular disease, reducing the autoimmune damage while leaving more of the immune system intact. This shift toward precisely targeted therapies is one of the more hopeful developments for people living with these conditions, though a true cure, one that restores tolerance to the specific self-target without suppressing everything, remains a goal researchers are still working toward.

Allergies and Hypersensitivity Reactions

An allergy is, at its core, a mistake of proportion. Your immune system encounters something genuinely harmless, a bit of pollen, a peanut protein, a cat's dander, and treats it as a dangerous invader, mounting a full-scale response against a threat that was never there. The pollen doesn't hurt you. Your reaction to the pollen does. Allergies and related overreactions belong to a broader category immunologists call hypersensitivity, meaning an immune response that's excessive or inappropriate and ends up damaging your own body.

Scientists classify hypersensitivity reactions into four types, a scheme developed by two immunologists, Robin Coombs and Philip Gell, in the 1960s that's still used today. The four types differ in which part of the immune system causes the damage and how quickly the reaction happens. Getting a feel for all four is worth the effort, because they explain a surprising range of conditions, and because the first type is what most people mean by "allergy."

Type I hypersensitivity is the immediate allergic reaction, and it's driven by the IgE antibody you met earlier. This is the classic allergy: hay fever, food allergies, allergic asthma, and the reaction to a bee sting. It happens fast, within minutes of exposure, which is why an allergic reaction comes on so suddenly. Because type I reactions cause the great majority of what people experience as allergies, they deserve the closest look, so let's walk through how one actually unfolds.

A type I allergy develops in two stages, and the first stage is silent. The very first time you encounter an allergen, you don't have a reaction. Instead, your immune system mistakenly identifies the harmless substance as a threat and produces IgE antibodies against it, a process called sensitization. These IgE antibodies then attach to the surface of mast cells, the tripwire cells stationed throughout your tissues, arming them and leaving them loaded. Nothing happens yet, and you have no idea you've become sensitized. But now you're primed. The next time you encounter that same allergen, it binds to the IgE waiting on your mast cells, and this triggers the mast cells to burst open and dump their contents, releasing a flood of histamine and other inflammatory chemicals all at once. That sudden release causes everything you recognize as an allergic reaction. This is why a person can eat a food or get stung many times with no problem and then suddenly develop an allergy: the earlier exposures were sensitizing them silently, and the reaction only appears once they're primed.

The specific symptoms of a type I reaction depend on where in the body it happens, and they all trace back to that histamine release. In your nose and eyes, it causes the sneezing, runny nose, and itchy watery eyes of hay fever. In your airways, it causes the constriction and swelling of an asthma attack, making it hard to breathe. In your skin, it causes hives, welts, and itching. In your gut, from a food allergy, it can cause cramping, vomiting, and diarrhea. Same underlying mechanism, different location, different symptoms. This is also exactly why antihistamine medications relieve so many allergy symptoms: they block the histamine that mast cells release, interrupting the reaction. When you take an antihistamine for seasonal allergies, you're blocking this precise step.

The most dangerous form of a type I reaction is anaphylaxis, and it's a genuine emergency worth understanding clearly. When the allergen enters the bloodstream and mast cells throughout the entire body release their contents at once, the result is a massive, body-wide reaction. Blood vessels everywhere dilate and leak, causing blood pressure to plummet, while the airways swell and constrict, making breathing difficult or impossible. Without rapid treatment, anaphylaxis can kill within minutes. Severe allergies to foods like peanuts and shellfish, to insect stings, and to certain drugs can all trigger it. The treatment is epinephrine, also called adrenaline, which rapidly reverses the reaction by tightening blood vessels and opening the airways, which is why people with severe allergies carry an epinephrine auto-injector, commonly known by the brand name EpiPen. That single injection can be the difference between life and death, and it works by directly counteracting the effects of the mast cell chemicals on the blood vessels and airways.

The other three types of hypersensitivity are worth knowing, even briefly, because they explain conditions that don't fit the classic allergy pattern and they connect back to mechanisms from earlier chapters.

Type II hypersensitivity is caused by antibodies, specifically IgG or IgM, binding directly to cells, marking those cells for destruction. Here the antibodies attack the body's own cells or transfused foreign cells. A mismatched blood transfusion causes a type II reaction, as your antibodies attack the incompatible red cells. Hemolytic disease of the newborn, where a mother's antibodies cross the placenta and attack her baby's red blood cells over a blood type incompatibility, is another example. Some autoimmune diseases work through this mechanism too, which shows that the line between allergy and autoimmunity isn't always sharp; both can involve the same underlying processes aimed at different targets.

Type III hypersensitivity is caused by immune complexes, the clumps that form when antibodies bind their target and stick together. Normally these complexes get cleared away, but when too many form, they can deposit in tissues like the walls of blood vessels, the kidneys, and the joints, where they trigger inflammation and damage. This is the mechanism behind some of the damage in lupus, as mentioned earlier, and behind a reaction called serum sickness. The problem in type III isn't the antibody attacking a cell directly, but the physical deposit of antibody-antigen clumps causing inflammation wherever they lodge.

Type IV hypersensitivity is the odd one out, because it isn't caused by antibodies at all. It's caused by T cells, and it's also called delayed-type hypersensitivity because it takes longer to develop, usually a day or two after exposure rather than minutes. Here, T cells respond to an antigen and drive inflammation directly. The itchy, blistering rash of poison ivy is a type IV reaction: a small molecule from the plant binds to proteins in your skin, and T cells attack, but the reaction takes a day or so to appear, which is why the rash shows up well after you touched the plant. The tuberculosis skin test works on this principle too: a small amount of TB protein is injected under the skin, and if you've been exposed to TB, your T cells react over the next two days, producing a firm bump that the doctor measures. The delay is the signature of a T cell response rather than an antibody one.

Type	Cause	Timing	Examples
Type I	IgE and mast cells	Immediate (minutes)	Hay fever, food allergy, anaphylaxis
Type II	IgG/IgM against cells	Hours	Transfusion reactions, hemolytic disease of newborn
Type III	Immune complexes deposit	Hours to days	Lupus damage, serum sickness
Type IV	T cells	Delayed (days)	Poison ivy, TB skin test, contact dermatitis

A natural question hangs over this whole subject: why do allergies exist at all? Why would evolution build a system that reacts so violently to harmless pollen and peanuts? One leading idea is the hygiene hypothesis, first proposed by a researcher named David Strachan in 1989 after he noticed that children with more older siblings had fewer allergies. The idea is that our immune systems evolved expecting heavy exposure to microbes, parasites, and dirt early in life, and that this exposure helps train the system to respond appropriately, especially teaching it not to overreact. In modern, cleaner environments, with fewer infections and less exposure to the microbes and parasites our ancestors dealt with, the immune system may be under-trained and more prone to misfiring against harmless substances. This helps explain why allergies and asthma have become much more common in wealthy, urbanized, hygienic societies. There's another layer to it: the IgE and Th2 machinery behind allergies evolved largely to fight parasitic worms, as you saw in the last chapter. In places where those parasites have been eliminated, that whole system may be left without its intended target, and it seems more likely to turn against innocuous things instead. Allergies may partly be the price of a defense that no longer has parasites to fight.

The hygiene hypothesis has practical consequences that overturned old medical advice. For years, parents were told to keep young children away from common allergens like peanuts to prevent allergies from developing. Research eventually showed the opposite: introducing peanuts to infants early, rather than avoiding them, actually reduces the chance of developing a peanut allergy, because early exposure helps the immune system learn to tolerate the food. A landmark study published in 2015 demonstrated this clearly, and medical guidance changed as a result. It's a good example of how understanding the immune mechanism, in this case how tolerance develops, can flip conventional wisdom on its head.

Immunodeficiency Disorders

The dysfunctions we've covered so far involve an immune system doing too much, attacking the wrong things or overreacting. Immunodeficiency is the opposite problem: an immune system that does too little. When part of the immune system is missing or broken, the body can't defend itself properly, and infections that a healthy person would shrug off become serious or even fatal. The hallmark of immunodeficiency is infections that are unusually frequent, unusually severe, unusually prolonged, or caused by microbes that don't normally trouble healthy people.

Immunodeficiencies come in two kinds, and the distinction matters for both cause and treatment. Primary immunodeficiencies are present from birth, caused by genetic defects in some part of the immune system. Secondary immunodeficiencies, also called acquired, develop later in life due to some outside factor that damages a previously healthy immune system. Primary immunodeficiencies are relatively rare and are essentially manufacturing defects in the immune system; secondary immunodeficiencies are far more common and can affect anyone.

The primary immunodeficiencies are a large group, because so many different genes are involved in building a working immune system, and a defect in any one of them can cause a specific deficiency. The pattern of infections often reveals which part is broken, which is one reason studying these conditions has taught immunologists so much. A defect in antibody production leads to repeated bacterial infections, since antibodies are central to clearing bacteria. A defect in T cells leads to trouble with viruses and fungi. A defect in phagocytes leads to problems clearing certain bacteria and fungi. The most severe primary immunodeficiency is severe combined immunodeficiency, or SCID, in which both the T cell and B cell arms of adaptive immunity are missing or nonfunctional. A baby born with SCID has almost no working adaptive immunity and is vulnerable to overwhelming infection from birth. This is the condition behind the famous case of David Vetter, the "bubble boy," who lived his entire short life in a sterile plastic enclosure in the 1970s and early 1980s to keep him away from germs, because his body couldn't fight even ordinary microbes. He died in 1984 after an attempted bone marrow transplant. Today, SCID can often be treated, and even cured, with a bone marrow transplant that gives the child a working immune system from a donor, and in some cases with gene therapy that corrects the underlying genetic defect. Many places now screen newborns for SCID so it can be caught and treated before a serious infection takes hold. Not all primary immunodeficiencies are so severe; the most common one, selective IgA deficiency, affects roughly one in several hundred people, and many of them have few or no symptoms because their other antibody classes largely compensate.

Secondary immunodeficiencies are much more common, and the causes are varied. The most well-known worldwide is HIV infection, which, as you learned in the last chapter, destroys helper T cells and, if untreated, progresses to AIDS, leaving the person open to the opportunistic infections and cancers that define the disease. But HIV is far from the only cause. Malnutrition is actually the most common cause of immunodeficiency globally, because building and running an immune system requires adequate protein and nutrients, and severe undernutrition weakens immune defenses, which is a major reason infections are so deadly in places with widespread hunger. Medical treatments are another significant cause: chemotherapy for cancer kills rapidly dividing cells, including the bone marrow cells that produce immune cells, so cancer patients often have dangerously low immune cell counts and are highly vulnerable to infection during treatment. The immunosuppressive drugs given to autoimmune patients and organ transplant recipients deliberately weaken immunity, which is the point of the treatment, but it leaves those patients more prone to infections as a side effect. Even ordinary aging weakens the immune system, a gradual decline called immunosenescence, which is a large part of why older adults are more susceptible to infections and why illnesses like influenza and pneumonia are more dangerous for the elderly.

Feature	Primary Immunodeficiency	Secondary Immunodeficiency
Cause	Genetic defect, present at birth	Outside factor damaging normal immunity
Frequency	Relatively rare	Common
Examples	SCID, selective IgA deficiency	HIV/AIDS, malnutrition, chemotherapy, aging
Onset	From birth or early life	Any time, depending on cause

Studying immunodeficiency has done something unexpected for the whole field: it's revealed exactly what each part of the immune system is for. When a particular component is missing and you see precisely which infections take hold as a result, you learn what that component was protecting against. Much of our understanding of the normal immune system was worked out by watching what goes wrong when specific pieces are absent, which is one of the reasons these rare conditions have mattered far beyond the patients they affect.

Chronic Inflammation and Disease Risk

You already know that inflammation is a defense, and that a healthy inflammatory response flares up to fight a threat and then resolves. Chronic inflammation is what happens when that shutdown fails and the response smolders on, at a low level, for months or years, in tissues that should be at rest. This lingering, unresolved inflammation has emerged as one of the more important ideas in modern medicine, because it's now linked to a startling number of the diseases that kill people in wealthy countries.

The difference between acute and chronic inflammation is a bit like the difference between a fire that burns briefly and is put out versus a fire that never quite goes out and slowly chars everything around it. Acute inflammation is intense but brief and beneficial. Chronic inflammation is low-grade but persistent, and over time it damages the very tissues it's occurring in. Instead of protecting you, it gradually wears you down. What keeps it going varies: sometimes a persistent infection the body can't fully clear, sometimes ongoing exposure to an irritant, sometimes an autoimmune process, and sometimes factors related to lifestyle and aging that keep the immune system in a mildly activated state.

The connection that has drawn the most attention is between chronic inflammation and heart disease, because it changed how we understand the leading cause of death in much of the world. Atherosclerosis, the buildup of fatty plaques in the walls of arteries that leads to heart attacks and strokes, was long thought of as simply a plumbing problem, cholesterol clogging pipes. We now understand it's substantially an inflammatory disease. Immune cells become involved in the artery wall, inflammation drives the growth and instability of the plaques, and it's often inflammation that causes a plaque to rupture and trigger the clot that actually causes a heart attack. This is why doctors sometimes measure a marker of inflammation in the blood, called C-reactive protein, to help assess a person's heart disease risk, and why inflammation has become a target of interest in preventing cardiovascular disease.

The list of conditions linked to chronic inflammation goes well beyond the heart. Type 2 diabetes involves chronic inflammation that interferes with the body's response to insulin. Chronic inflammation appears to contribute to the development of some cancers, since a constantly inflamed environment can promote the growth and mutation of cells, which is part of why certain long-term infections and inflammatory conditions raise cancer risk. Inflammation is increasingly implicated in neurodegenerative diseases like Alzheimer's, where inflammation in the brain seems to contribute to the damage. And a great deal of the general decline associated with aging is now thought to involve chronic low-grade inflammation, a phenomenon researchers have nicknamed "inflammaging," the slow, persistent rise in background inflammation as we get older.

What drives this widespread chronic inflammation? Several factors related to modern life appear to contribute, and this is where the topic becomes personally relevant. Obesity is a major one: fat tissue, especially around the abdomen, isn't just inert storage but actively produces inflammatory signals, so carrying excess fat keeps the body in a mildly inflamed state. Diet plays a role, with certain eating patterns promoting inflammation and others reducing it. Chronic stress, poor sleep, smoking, and a sedentary lifestyle all appear to push inflammation upward. This connects the immune system directly to the everyday factors that shape long-term health, and it's part of why the same lifestyle habits keep coming up in the prevention of so many different diseases: many of them share chronic inflammation as a common thread. The immune system, it turns out, sits quietly in the background of some of the most common causes of death, not through dramatic infections but through this slow, persistent, low-level activation.

Cancer and Immune Surveillance

Cancer might seem out of place in a book about immunity, but it belongs here for a reason you first encountered in the opening chapter: one of your immune system's ongoing jobs is to find and destroy your own cells when they turn dangerous. Cancer is, in a sense, a failure of that surveillance, and understanding the relationship between cancer and the immune system has transformed cancer treatment in recent years.

Cancer begins when one of your own cells accumulates mutations that let it grow and divide without the normal controls, ignoring the signals that tell it when to stop. Because a cancer cell comes from your own body, it's a peculiar kind of enemy: mostly self, but dangerously altered. The idea that the immune system watches for and eliminates these cells before they can form tumors is called immune surveillance, an idea proposed decades ago by scientists including Frank Macfarlane Burnet, whom you met earlier in connection with clonal selection. The theory holds that cancer cells arise more often than we realize, and that the immune system routinely catches and destroys most of them, so that the cancers that actually develop are the rare ones that slipped past.

How can the immune system recognize a cancer cell when it started as one of your own? The answer is that cancer cells aren't quite normal anymore. As they mutate, they often produce abnormal proteins, or produce normal proteins in abnormal amounts, and fragments of these altered proteins get displayed on the cell's MHC class I molecules, the same display system that reveals viral infections. Cytotoxic T cells and natural killer cells can recognize these abnormal displays as signs that a cell has gone wrong and kill it, the same way they'd kill a virus-infected cell. So the tools your body uses against cancer are ones you already know: the cytotoxic T cells and NK cells that patrol for cells displaying something abnormal.

But cancers that grow into actual disease have, by definition, escaped this surveillance, and the story of how they do it is one of the more sobering examples of natural selection at work inside a single body. Immunologists describe the process in three phases, sometimes called the three E's: elimination, equilibrium, and escape. In the elimination phase, the immune system successfully finds and destroys cancer cells as they arise, and surveillance works as intended. But if some cancer cells survive, the process enters equilibrium, a prolonged standoff in which the immune system keeps the cancer in check without fully eliminating it, sometimes for years. Here's the dangerous part: during this standoff, the immune system is unintentionally acting as a selective pressure, killing off the cancer cells it can recognize and leaving behind the rare variants that have acquired ways to hide. Over time, the surviving cancer cells become better and better at evading the immune system, until finally, in the escape phase, a variant emerges that the immune system can no longer control, and the cancer grows into detectable disease. The immune system, by attacking the cancer, inadvertently trained it to become invisible. This grim process is called immunoediting, because the immune system edits the cancer, selecting for the cells best able to escape it.

Phase	What Happens	Outcome
Elimination	Immune system finds and destroys cancer cells	Cancer prevented; surveillance succeeds
Equilibrium	Immune system holds cancer in check but can't clear it	Standoff; resistant variants selected over time
Escape	A variant evades immune control and grows	Cancer becomes detectable disease

The tricks cancers use to escape are worth understanding, because they're exactly what modern treatments try to counteract. Some cancer cells reduce or eliminate the MHC molecules on their surface, hiding the abnormal fragments that would flag them to T cells, though this can make them targets for NK cells instead. Many tumors create an immunosuppressive environment around themselves, releasing signals that shut down nearby immune cells and recruiting regulatory T cells, the immune system's own brakes, to protect the tumor from attack. It's a striking betrayal: the tumor hijacks the very cells meant to prevent excessive immune responses and uses them as a shield. And crucially, many cancers exploit the immune system's natural "off switches," molecules called checkpoints that normally prevent the immune system from overreacting. Cancer cells can display signals that engage these checkpoints, effectively telling attacking T cells to stand down, and the T cells obey. The tumor waves a flag that says "I'm friendly, don't attack," and the immune system's own safety mechanisms cause it to comply.

This last discovery, that cancers escape by pulling the immune system's own brakes, led to one of the most important advances in cancer treatment in decades. If tumors survive by engaging these checkpoints to shut down T cells, then a drug that blocks the checkpoints could release the brakes and let the immune system attack the cancer freely. That's exactly what a class of drugs called checkpoint inhibitors does, and they've produced remarkable results in cancers that were once nearly untreatable. This is only one part of a broader effort to turn the immune system against cancer deliberately, an approach called immunotherapy that has become one of the most active frontiers in all of medicine. Because it's such a large and rapidly developing subject, and because it builds on everything in this book, we've given it a full treatment in the final chapter. For now, the connection to hold onto is that cancer and immunity are deeply intertwined: your immune system is your first defense against cancer, cancers succeed largely by evading that defense, and the newest treatments work by restoring the immune system's ability to do the job it was always meant to do.

Discussion Question

In autoimmune disease the immune system attacks the body, and in cancer it fails to attack cells that have become dangerous. Both are failures of the same underlying task: correctly deciding what counts as a threat. Given that making this distinction perfectly may be impossible, do you think an immune system tuned to be more aggressive, catching more cancers but causing more autoimmunity, would be better or worse than one tuned the other way, and what does that tell you about why evolution settled where it did?

Chapter 8: Vaccination and Immunization

Here's a number worth sitting with: vaccines prevent several million deaths every single year. No drug, no surgery, no other single medical tool comes close to that record. And they manage it through the simplest idea in this entire book, one you've already met more than once. Show the immune system a harmless preview of an enemy, and it will remember.

This chapter is about how that preview gets built and put to use. We'll start with what a vaccine actually does inside you, then work through the different ways vaccines are made, from methods that are more than two centuries old to the mRNA technology that arrived in force only recently. After that we'll look at how vaccinating enough people protects even those who aren't vaccinated, how vaccines are tested for safety, and how the world has used them to push some diseases to the very edge of extinction.

How Vaccines Train the Immune System

You already understand the core principle from the chapters on adaptive immunity, so let's build on it rather than repeat it. A vaccine works by producing memory: after a safe exposure to some antigen, your body keeps memory T cells and B cells and a supply of refined antibodies, so that a later encounter with the real pathogen triggers the fast, overwhelming secondary response instead of the slow, painful first one. That much you know. What this section does is look at how vaccines actually accomplish this in practice, what goes into them, and why they differ so much from one another.

Start with the history, because it explains why the whole field is even called vaccination. Long before anyone understood immunity, people noticed that surviving certain diseases protected you from getting them again. In the case of smallpox, one of the great killers of human history, this observation led to a risky practice called variolation: deliberately infecting a healthy person with material from a mild smallpox case, in the hope they'd get a milder illness and come out immune. Variolation was practiced for centuries in China, India, and the Ottoman world, and it was brought to England in the early 1700s by Lady Mary Wortley Montagu, who had seen it in Constantinople. It worked, but it was dangerous, because you were giving people actual smallpox, and some of them died from it.

The safer breakthrough came in 1796, from an English country doctor named Edward Jenner. Jenner acted on a piece of local folk knowledge: milkmaids who caught cowpox, a mild disease they got from cows, seemed not to catch smallpox. Cowpox is closely related to smallpox but far milder in humans. Jenner took material from a cowpox sore on a milkmaid's hand and used it to inoculate an eight-year-old boy named James Phipps. When he later exposed the boy to smallpox, the boy didn't get sick. Jenner had shown that a harmless relative of a dangerous disease could produce protection against the dangerous one, with far less risk than variolation. He named the technique after the Latin word for cow, vacca, which is why we call it vaccination to this day.

For nearly a century, vaccination meant only smallpox. Then Louis Pasteur, working in France in the 1880s, generalized the idea and turned it into a science. Pasteur discovered, partly by accident, that if you weakened a pathogen by growing it under unusual conditions, you could use the weakened version to protect against the deadly one. He developed vaccines against chicken cholera, anthrax in livestock, and most famously rabies, which he first used to save a boy named Joseph Meister who had been bitten by a rabid dog in 1885. Pasteur deliberately kept Jenner's word "vaccine" for all of these, as a tribute, even though none of them involved cows. From that point on, vaccination became a general strategy that could be applied to disease after disease.

So what's actually inside a vaccine? The main ingredient is the antigen: the piece of the pathogen, or the weakened pathogen itself, that your immune system learns to recognize. Everything else in the vaccine is there to support that antigen or to keep the product safe and stable. Many vaccines contain an adjuvant, which you met in the last chapter, a substance that heightens the immune response so a small amount of antigen produces strong, lasting immunity. Aluminum salts have been used as adjuvants for decades and remain the most common. Some vaccines also contain small amounts of stabilizers to protect the antigen, and preservatives to keep the vial from becoming contaminated once it's opened. The amounts are tiny, and each component has a specific job.

There's a concept that helps make sense of why different vaccines are designed so differently, and it's worth knowing: the correlate of protection. This is the measurable immune response that tells you a vaccine is working. For many vaccines, protection correlates with a certain level of neutralizing antibodies in the blood, so vaccine developers aim to produce at least that level. But for some diseases, antibody levels alone don't tell the whole story, and T cell responses matter more. Knowing what actually protects against a given disease, whether it's antibodies, T cells, or both, shapes how the vaccine is built and how doctors judge whether it's doing its job.

This also explains why some vaccines need booster shots and others don't. A single vaccine, or a short initial series, sometimes produces immunity that lasts a lifetime, as with the measles vaccine. Other vaccines produce immunity that fades over years, so a booster is needed to reactivate the memory cells and push antibody levels back up. The tetanus vaccine is a familiar example: protection wanes over roughly a decade, which is why adults are advised to get a tetanus booster every ten years. Whether a vaccine needs boosters depends on the disease, the type of vaccine, and how durable the memory it produces turns out to be. There's a general pattern worth carrying into the next section: vaccines made from living, replicating microbes tend to produce the strongest and longest-lasting immunity, while vaccines made from killed microbes or isolated pieces tend to produce weaker immunity that often needs boosting. That difference is the reason there are so many kinds of vaccine in the first place.

Live-Attenuated vs. Inactivated Vaccines

The oldest and most fundamental split in vaccine design is between using a living pathogen that's been weakened and using one that's been killed outright. These two approaches, live-attenuated and inactivated, represent a genuine trade-off between how strong the immunity is and how safe the vaccine is to give, and understanding that trade-off is the doorway to understanding every other vaccine type.

A live-attenuated vaccine uses a living form of the pathogen that's been weakened, or attenuated, so it can still replicate in your body but can no longer cause disease. This is essentially what Jenner and Pasteur were doing. The weakened microbe multiplies inside you just enough to provoke a full immune response, complete with the whole machinery of infection, but without making you sick. Because the immune system responds to a live, replicating microbe much as it would to a real infection, the immunity produced is strong, broad, and long-lasting, often for life after just one or two doses. Live-attenuated vaccines also tend to stimulate both antibody responses and strong T cell responses, giving well-rounded protection.

Many of the most successful vaccines are live-attenuated. The MMR vaccine, which protects against measles, mumps, and rubella in a single shot, uses weakened versions of all three viruses and produces immunity that usually lasts a lifetime. The chickenpox (varicella) vaccine, the oral polio vaccine developed by Albert Sabin, the yellow fever vaccine, the rotavirus vaccine given to infants, and the BCG vaccine against tuberculosis are all live-attenuated. The nasal spray flu vaccine uses a weakened live virus too. These vaccines have prevented enormous amounts of disease, and their strength is exactly what makes them so effective.

But that strength comes with real limitations, and they matter. Because a live-attenuated vaccine contains a replicating microbe, it can be dangerous for people whose immune systems are weakened, the very immunocompromised patients from the last chapter. In a person with a healthy immune system, the weakened microbe is easily controlled, but in someone with a failing immune system, even a weakened microbe can potentially cause illness. This is why live-attenuated vaccines generally can't be given to people with severe immunodeficiency, to patients on strong immunosuppressive drugs, or usually to pregnant women, out of caution for the developing baby. Live vaccines also need careful handling: they require refrigeration to keep the microbe alive and viable, which creates logistical challenges we'll return to when we discuss global programs. And there's a small but real risk that a weakened microbe could mutate back toward a more dangerous form. This has happened rarely with the oral polio vaccine, where the weakened virus can, in very rare cases, revert and cause vaccine-derived polio, a problem serious enough that it has shaped global polio strategy.

An inactivated vaccine takes the opposite approach: it uses a pathogen that's been completely killed, usually with heat or chemicals, so it can't replicate or cause disease at all. The dead microbe still carries its antigens, so your immune system can still recognize it and respond, but since nothing is replicating, there's no infection of any kind. This makes inactivated vaccines much safer in one clear way: because the microbe is dead, it can't cause the disease and it can't revert to a dangerous form, so these vaccines can be given safely to immunocompromised people and pregnant women. Inactivated vaccines are also more stable and easier to store than live ones, since keeping something alive is no longer a concern.

The most famous inactivated vaccine is the injected polio vaccine developed by Jonas Salk, which uses killed poliovirus and is the version used in most wealthy countries today. Most flu shots are inactivated. The hepatitis A vaccine and the rabies vaccine given to people after exposure are inactivated too. These vaccines are safe and reliable, but they pay for that safety with weaker immunity. Because a dead microbe doesn't replicate and doesn't fully mimic a real infection, the immune response it provokes is generally weaker and shorter-lived than what a live vaccine produces. This is why inactivated vaccines usually require multiple doses to build up protection and often need periodic boosters to maintain it, and why they frequently include an adjuvant to strengthen the response.

Here's the trade-off laid out directly.

Feature	Live-Attenuated	Inactivated
What it contains	Weakened but living pathogen	Completely killed pathogen
Strength of immunity	Strong, often lifelong	Weaker, shorter-lived
Doses needed	Often one or two	Usually several, plus boosters
Safe for immunocompromised	Generally no	Yes
Storage	Needs refrigeration; more fragile	More stable
Risk of reverting	Very small but possible	None; microbe is dead
Examples	MMR, chickenpox, oral polio, yellow fever, BCG	Injected polio, hepatitis A, most flu shots, rabies

Live and inactivated whole-pathogen vaccines are the two classic approaches, but they aren't the only ones. Over the past several decades, scientists have developed vaccines that use only pieces of a pathogen rather than the whole thing, an approach that can be even safer because there's no chance of causing the disease when the disease-causing parts aren't even present. These are worth walking through, because several of them protect against illnesses you'll recognize.

A subunit vaccine, sometimes called a protein subunit vaccine, uses only specific pieces of the pathogen, usually one or a few of its proteins, rather than the entire microbe. Your immune system learns to recognize those particular proteins, and since the rest of the pathogen isn't included, there's nothing that could cause infection. The hepatitis B vaccine is a landmark example: it uses a single surface protein of the hepatitis B virus, and it was one of the first vaccines made through genetic engineering, with yeast cells programmed to manufacture the viral protein. The HPV vaccine, which protects against the human papillomavirus that causes cervical and other cancers, uses viral proteins that assemble into empty shells resembling the virus, called virus-like particles, which look like the virus to the immune system but contain no viral genetic material and can't infect anything. The acellular pertussis vaccine, part of the childhood whooping cough shot, uses purified pieces of the pertussis bacterium rather than the whole cell. And the newer shingles vaccine combines a viral protein with a strong adjuvant to produce excellent protection in older adults. Subunit vaccines are very safe, but like inactivated vaccines, they often need adjuvants and boosters to produce strong immunity.

A toxoid vaccine is a clever variation used for diseases where the danger comes from a bacterial toxin rather than the bacterium itself, a situation you met in the chapter on infection. A toxoid is a toxin that's been inactivated so it can no longer harm you but can still provoke an immune response. Your body learns to make antitoxin antibodies that neutralize the real toxin if you're ever exposed. The tetanus and diphtheria vaccines are toxoid vaccines, and they're the reason these once-feared diseases are now rare in vaccinated populations. You're not being protected against the bacterium so much as against the poison it produces.

A conjugate vaccine solves a specific problem that stumped vaccine developers for years, and it connects directly to something from the chapter on adaptive immunity. Recall that certain bacteria are coated in sugar capsules, and that these sugar antigens can activate B cells directly without T cell help, in a T-independent response that's weak and produces poor memory, and that young children under about two mount especially poor responses to these sugars. This meant that plain sugar-based vaccines couldn't protect the infants who needed them most against dangerous capsule-covered bacteria. The solution was to chemically link, or conjugate, the bacterial sugar to a carrier protein. That linkage recruits T cell help into the response, converting a weak T-independent reaction into a strong, memory-forming one that works even in infants.

The Hib vaccine, against a bacterium called Haemophilus influenzae type b that once caused a great deal of childhood meningitis, was one of the first conjugate vaccines, and it nearly eliminated Hib meningitis in countries that adopted it. The pneumococcal conjugate vaccine and certain meningitis vaccines work the same way. Conjugate vaccines are one of the clearest cases of understanding the immune mechanism, in this case why infants respond poorly to sugars, and using that understanding to solve a real medical problem.

A viral vector vaccine uses a harmless virus as a delivery vehicle to carry genetic instructions for a target antigen into your cells. Scientists take a virus that doesn't cause disease, often a weakened adenovirus, and engineer it to carry the gene for a protein from the pathogen you actually want to protect against. When this harmless carrier enters your cells, your cells read the gene and produce the target antigen, and your immune system responds to it. Some of the COVID-19 vaccines used this approach, engineering an adenovirus to carry the gene for the coronavirus spike protein. The first approved Ebola vaccine also uses a viral vector, and it's been used to control Ebola outbreaks. Viral vector vaccines can produce strong responses, but the approach is more complex, and sometimes prior immunity to the carrier virus can interfere with the vaccine's effect.

Here's a summary of the main vaccine platforms and how they compare.

Platform	How It Works	Examples	Notes
Live-attenuated	Weakened living pathogen replicates and provokes strong immunity	MMR, chickenpox, oral polio, yellow fever	Strongest immunity; not for immunocompromised
Inactivated	Killed pathogen provokes a response without replicating	Injected polio, hepatitis A, most flu shots	Safe but weaker; needs boosters
Subunit	Only pieces of the pathogen are used	Hepatitis B, HPV, shingles, acellular pertussis	Very safe; often needs adjuvant
Toxoid	Inactivated toxin trains antitoxin antibodies	Tetanus, diphtheria	Protects against the toxin, not the microbe
Conjugate	Bacterial sugar linked to a protein to engage T cells	Hib, pneumococcal, meningococcal	Works in infants where plain sugars fail
Viral vector	Harmless virus delivers a gene for the target antigen	Some COVID-19 vaccines, Ebola	Cells make the antigen; more complex

Look across the platforms and a pattern emerges. The whole history of vaccine development has moved toward using less and less of the actual pathogen: from the whole living microbe, to the whole killed microbe, to isolated pieces, to just the genetic instructions for a piece.

Each step generally increases safety, because there's less that could go wrong, though it often requires more cleverness to produce a strong enough response. That trajectory leads directly to the newest platform, which takes the idea to its logical end point by delivering nothing but genetic instructions.

Amount of pathogen material

	Live-attenuated	Inactivated	Subunit / toxoid	Viral vector	mRNA
Antibody response	● ● ●	● ● ○	● ● ○	● ● ●	● ● ●
T-cell response	● ● ●	● ○ ○	● ○ ○	● ● ●	● ● ●
Boosters needed	● ○ ○	● ● ●	● ● ●	● ● ○	● ● ○
Safe if immune-weak	● ○ ○	● ● ●	● ● ●	● ● ○	● ● ●
Storage demands	● ● ●	● ○ ○	● ○ ○	● ● ○	● ● ●
Reversion risk	● ● ○	● ○ ○	● ○ ○	● ○ ○	● ○ ○

● filled = higher ○ open = lower

Less pathogen, more control; no platform wins on every count

mRNA Vaccine Technology

The mRNA vaccine represents the most dramatic shift in vaccine design in a generation, and its sudden arrival during the COVID-19 pandemic made it one of the most talked-about pieces of medical technology in the world. To understand why it's such a departure, and why so many scientists are excited about it, you need to grasp the elegant idea at its center: instead of putting an antigen into your body, an mRNA vaccine gives your own cells the instructions to build the antigen themselves.

To follow how this works, recall a piece of basic biology. Your genes, written in DNA, contain the instructions for making all your proteins. When a cell needs to make a particular protein, it first copies the relevant gene into a temporary working instruction called messenger RNA, or mRNA. This mRNA carries the instruction out to the cell's protein-building machinery, the ribosomes, which read it and assemble the protein accordingly. The mRNA is a short-lived, disposable message: once the protein is made, the mRNA is quickly broken down. It's the cell's normal way of turning genetic instructions into actual proteins, and it happens constantly in every cell you have.

An mRNA vaccine takes advantage of this everyday process. Scientists design a piece of mRNA carrying the instructions for one specific protein from a pathogen, an antigen, such as the spike protein that studs the surface of the coronavirus. This mRNA is injected into your arm, where it enters some of your cells. Your cells' own ribosomes read the mRNA and manufacture the antigen, exactly as they would any protein. The antigen then appears on and around your cells, and your immune system, seeing this foreign protein, responds to it, producing antibodies and memory cells against it. In effect, your cells briefly become tiny factories producing the target antigen, and your immune system learns to recognize it. You never receive the pathogen itself, only the instructions for making one harmless piece of it.

A common worry about this technology deserves a clear and honest answer, because it comes up constantly. Many people fear that an mRNA vaccine could alter their DNA. It cannot, and the reason lies in basic cell biology. The mRNA never enters the nucleus of your cells, where your DNA is stored, and it has no ability to insert itself into your genome. mRNA and DNA are different molecules, and the machinery to convert mRNA back into DNA and splice it into your chromosomes simply isn't present in the normal process. Moreover, the mRNA is inherently short-lived: like all mRNA in your cells, it's broken down within a matter of hours to days after doing its job, and then it's gone. It delivers its instruction, your cells make the antigen for a little while, and the mRNA degrades. It leaves no lasting genetic trace whatsoever.

There were two hard technical problems that had to be solved before mRNA vaccines could work, and solving them took decades of patient research. The first problem is that mRNA is extremely fragile. On its own, injected mRNA would be torn apart almost instantly by enzymes in your body before it ever reached a cell. The solution was to wrap the mRNA in a protective bubble of fat molecules called a lipid nanoparticle. These tiny fatty spheres shield the mRNA from destruction and help carry it into your cells, where it can be released to do its work. Without lipid nanoparticles, the mRNA would never survive the journey.

The second problem was subtler and took longer to crack. Your body's innate immune system, with its pattern recognition receptors, treats foreign RNA as a danger signal, since viral RNA is exactly the kind of thing those receptors evolved to detect. Early attempts at mRNA therapies triggered so much inflammation that they were unusable, and the immune reaction also shut down the production of the desired protein. The breakthrough came from two researchers, Katalin Karikó and Drew Weissman, who spent years working on this problem, often against considerable skepticism and funding difficulties. They discovered that by subtly modifying one of the chemical building blocks of the mRNA, they could make it far less inflammatory while actually increasing how much protein the cells produced from it. This modification made mRNA vaccines practical, and it's used in the vaccines that reached the public. Karikó and Weissman received the Nobel Prize in Physiology or Medicine in 2023 for this work, an award that recognized research that had struggled for support for much of its history before proving essential during a global emergency.

The advantages of mRNA vaccines are considerable, and they explain the enthusiasm. The biggest is speed of design. To make a traditional vaccine, you generally have to grow the pathogen, or produce its proteins, which is slow and specific to each pathogen. To make an mRNA vaccine, you only need the genetic sequence of the target antigen, which can be determined from the pathogen's genome. Once you have the sequence, designing the matching mRNA is fast, and the manufacturing process is largely the same regardless of which antigen you're targeting. This is why the first COVID-19 mRNA vaccines were designed within days of the coronavirus's genetic sequence being published in January 2020, and why vaccines were in people's arms roughly eleven months later, an unprecedented pace.

The same speed makes it relatively easy to update mRNA vaccines when a virus mutates, since you can simply swap in the sequence for the new variant. mRNA vaccines also tend to produce strong immune responses, including good T cell responses, and because your own cells make the antigen, no live pathogen is involved, so there's no risk of the vaccine causing the disease.

There are challenges too, and honesty about them matters. The lipid nanoparticles and the mRNA can be sensitive to temperature, so the first mRNA vaccines required storage at very cold temperatures, one of them at around minus seventy degrees Celsius, which created serious logistical difficulties for distribution, especially in places without specialized freezers. Newer formulations have become more stable and easier to store, but temperature sensitivity remains a consideration. mRNA vaccines also tend to cause noticeable short-term side effects in many people, a sore arm, fatigue, headache, and sometimes a mild fever for a day or two, which are signs of the immune system responding and generally pass quickly. And because the technology is relatively new in widespread use, ongoing monitoring continues, as it does for all vaccines.

What makes mRNA technology genuinely important is that it's a platform, adaptable to many different diseases, not a one-time solution to a single virus. Researchers are now developing mRNA vaccines against influenza, against the respiratory virus RSV, and against a range of other infections, aiming to use the platform's speed and flexibility. Some of the most interesting work applies mRNA to cancer: because a tumor's mutations produce abnormal proteins unique to that patient's cancer, scientists can, in principle, sequence a patient's tumor, identify those unique abnormal proteins, and design a personalized mRNA vaccine that trains the patient's immune system to attack their specific cancer. This kind of individualized cancer vaccine is in active clinical trials and represents one of the more promising directions the technology could take. The mRNA vaccine may end up being remembered less for any single vaccine than for opening a fast, flexible new way to instruct the immune system, which is a subject the final chapter returns to.

Herd Immunity

So far we've talked about vaccination as something that protects the individual who receives it. But vaccines do something more, something that only becomes visible when you look at a whole population, and it's one of the more beautiful ideas in public health. When enough people in a community are immune to a disease, the disease can't spread effectively, and it dies out, protecting even the people who aren't immune. This is called herd immunity, and it's the reason vaccination is as much a collective act as a personal one.

The logic is straightforward once you see it. A contagious disease spreads by passing from an infected person to susceptible people around them. If most of the people around an infected person are immune, the disease has nowhere to go: it hits mostly immune individuals who can't catch it or pass it on, and the chain of transmission breaks.

Even the few susceptible people in the community are protected, not because they're immune themselves, but because the disease can't reach them, since it keeps running into immune people and fizzling out. The immune majority forms a kind of barrier that shields the vulnerable minority.

This collective protection matters enormously because not everyone can be vaccinated. Newborns are often too young to have received certain vaccines. People with severe immunodeficiency or those on immunosuppressive treatment may not be able to take live vaccines, and may not respond well to any vaccine, as you saw in the last two chapters. Some people have genuine medical reasons, like a severe allergy to a vaccine component, that prevent vaccination. And even in vaccinated people, no vaccine is perfectly effective, so a fraction won't be fully protected. Herd immunity protects all of these vulnerable people, but only as long as enough of the surrounding population is immune. When a healthy person gets vaccinated, they're helping protect the infant, the cancer patient, and the elderly neighbor who can't be protected on their own.

How many people need to be immune for herd immunity to work? It depends on how contagious the disease is, and there's a way to calculate it. Epidemiologists measure contagiousness with a number called the basic reproduction number, written R-nought and pronounced "R-naught," which represents the average number of people that one infected person would spread the disease to in a population with no immunity. A higher R-nought means a more contagious disease, which means a higher fraction of the population must be immune to stop it. The threshold fraction can be calculated from the reproduction number, and the more contagious the disease, the closer to everyone you need.

The differences between diseases are striking. Measles is one of the most contagious diseases known, with a reproduction number often estimated between twelve and eighteen, meaning one case could infect a dozen or more susceptible people. This is why measles requires such a high immunity threshold, in the range of ninety-five percent, to prevent outbreaks. That's an extremely high bar, and it's why even small drops in measles vaccination rates can allow the disease to come roaring back, as has happened in communities where vaccination declined. A less contagious disease requires a lower threshold. Here's how a few diseases compare.

Disease	Approximate R-nought	Rough Immunity Threshold Needed
Measles	12 to 18	About 95%
Whooping cough (pertussis)	12 to 17	About 92 to 94%
Polio	5 to 7	About 80 to 86%
Smallpox	5 to 7	About 80 to 85%
Seasonal influenza	1 to 2	About 30 to 50%

These thresholds have real consequences, and the measles example shows what's at stake. When vaccination rates stay above the threshold, a disease can be held at bay or even eliminated from a region entirely. When rates fall below it, the barrier develops holes, and outbreaks return. Measles had been eliminated from many countries, meaning it no longer spread continuously within them, but declining vaccination in some communities has allowed outbreaks to reappear, sometimes traced to a single imported case that then spreads among the unvaccinated. Because measles is so contagious, it's often the first disease to return when vaccination rates slip, acting as an early warning that a community's immunity has dropped too low. Herd immunity, in other words, isn't a permanent achievement but a condition that has to be maintained, and it depends on enough people continuing to vaccinate.

There's an uncomfortable tension built into herd immunity that's worth naming honestly. Because the immune majority protects everyone, an individual might reason that they could skip a vaccine and still be protected by everyone else's immunity, getting the benefit without the small risks or effort of vaccination. If only a few people think this way, herd immunity holds. But if many do, the immune fraction drops below the threshold, the barrier fails, and the disease returns to harm the vulnerable. This is why vaccination is often treated as a shared responsibility rather than a purely individual choice: the protection it provides to those who can't be vaccinated only exists if enough people participate. Herd immunity is a genuinely collective good, produced by many individual decisions.

Vaccine Safety and Testing

Given how many people receive them, including healthy children, vaccines are held to an exceptionally high safety standard, higher in some ways than most medicines. A drug given to sick people is weighed against the harm of their disease. A vaccine is given to healthy people, often infants, to prevent a disease they don't yet have, so the tolerance for risk is very low, and the testing is correspondingly rigorous. Understanding how vaccines are tested and monitored is the best answer to the safety questions that many people reasonably have.

Every vaccine goes through a long series of steps before it reaches the public, and each step can stop a candidate that proves unsafe or ineffective. It begins with preclinical research, in the laboratory and in animals, to see whether the vaccine produces an immune response and shows no obvious safety problems before any human ever receives it. Only after this does testing move into people, and human testing proceeds through three phases of clinical trials, each larger than the last.

Phase I trials involve a small group, typically a few dozen healthy volunteers, and focus mainly on safety and on finding the right dose. Researchers watch closely for side effects and check whether the vaccine produces the intended immune response. Phase II trials expand to several hundred people and continue to assess safety and immune response while refining the dose and schedule, often including some of the groups the vaccine is ultimately meant for. Phase III trials are the largest and most demanding, involving thousands to tens of thousands of participants, and they're designed to prove that the vaccine actually works. In a phase III trial, volunteers are randomly assigned to receive either the vaccine or a placebo, and neither the participants nor the researchers know who got which until the trial is analyzed. The two groups are then followed to see whether the vaccinated group actually develops the disease less often than the placebo group. This randomized, blinded design is the gold standard for showing that a vaccine prevents disease and not merely that it produces antibodies. Phase III trials also catch less common side effects that smaller trials might miss, precisely because so many people are enrolled.

Phase	Number of Participants	Main Purpose
Preclinical	Lab and animal studies	Basic safety and immune response before human testing
Phase I	Tens	Safety and dosing in healthy volunteers
Phase II	Hundreds	Safety, immune response, dose refinement
Phase III	Thousands to tens of thousands	Prove real-world efficacy and catch rarer side effects

Only after a vaccine passes all three phases does it go to regulators, such as the Food and Drug Administration in the United States or the European Medicines Agency, who review all the data independently and decide whether to approve it. But approval isn't the end of the safety process, and this is a part many people don't realize exists. Once a vaccine is in widespread use, it continues to be monitored, in what's sometimes called phase IV or post-marketing surveillance. This ongoing monitoring matters because some side effects are so rare that they only become detectable once millions of people have been vaccinated, far more than any trial could include. Countries run systems to track this. In the United States, one system lets doctors and the public report any health problem that occurs after vaccination, and another links vaccination records with medical records across large populations to detect any unusual patterns. These systems are designed to catch even very rare problems and have done so, occasionally leading to changes in how a vaccine is used. The safety monitoring, in other words, continues for the entire life of a vaccine.

Most vaccine side effects are mild, expected, and actually signs that the vaccine is working. A sore arm, a low fever, tiredness, or a headache for a day or two after vaccination are common, and they reflect your immune system responding to the vaccine, mounting the very reaction that builds your protection. Serious side effects do occur, but they're genuinely rare, and the monitoring systems exist precisely to identify and quantify them so that the benefits and risks can be weighed honestly. For the diseases vaccines prevent, the risks of the disease itself are, in the overwhelming majority of cases, far greater than the risks of the vaccine.

No discussion of vaccine safety is complete without addressing the myth that vaccines cause autism, because it has caused real harm and it deserves a clear, factual response. The idea originated with a paper published in 1998 by a British doctor named Andrew Wakefield, which suggested a link between the MMR vaccine and autism. The paper was based on only twelve children, and it was later found to be not merely wrong but fraudulent: investigations revealed that Wakefield had manipulated the data, had undisclosed financial conflicts of interest, and had subjected children to unnecessary procedures. The paper was formally retracted by the journal in 2010, and Wakefield lost his license to practice medicine. In the years since, the question has been studied exhaustively, in large investigations involving millions of children across many countries, and these studies have consistently found no link whatsoever between vaccines and autism. The scientific evidence on this point is about as clear as evidence gets. Yet the original claim, despite being thoroughly discredited, contributed to a decline in vaccination in some communities and to outbreaks of preventable diseases like measles, which is a sobering example of how a single fraudulent study can do lasting damage to public health.

The rapid development of COVID-19 vaccines raised understandable questions about whether safety was compromised for speed, and the honest answer is that the speed came from specific factors, not from skipping steps. The mRNA platform had been in development for years beforehand, so the underlying technology wasn't invented from scratch. Extensive prior research on related coronaviruses, from the earlier SARS and MERS outbreaks, gave scientists a head start on which antigen to target. Enormous funding let manufacturers do things in parallel that are normally done one after another, such as beginning manufacturing before final approval, accepting the financial risk that the vaccine might fail. The trial phases were run with some overlap rather than sequentially, and the trials enrolled volunteers exceptionally quickly because the pandemic created both urgency and a large pool of willing participants. Critically, the actual clinical trials still enrolled tens of thousands of people and still had to demonstrate safety and efficacy through the normal randomized, blinded design, and regulators still reviewed the data. The steps that protect safety were followed; what changed was the removal of the usual delays between steps. Understanding this distinction, that speed came from parallel effort and prior groundwork rather than from cutting corners, is the key to understanding how the vaccines could be developed so fast while still being tested.

Global Vaccination Programs

The true measure of what vaccines can do shows up not in individual protection but in what happens when the world decides to use them at scale, deliberately and cooperatively, against a common enemy. The results have been among the great achievements of human civilization, and they're worth knowing not as a matter of pride but because they reveal both what's possible and what it takes.

The crowning example is the eradication of smallpox, the only human disease ever completely wiped from the Earth. Smallpox was one of history's most terrible killers, and even into the twentieth century it was killing millions of people a year and leaving many survivors blind or scarred. In 1967, the World Health Organization launched an intensified global campaign to eradicate it entirely, led in large part by an American epidemiologist named Donald Henderson. The campaign succeeded through a clever strategy: rather than trying to vaccinate every single person on the planet, which would have been impossible, teams used surveillance to find each outbreak and then vaccinated everyone around each case, a ring of contacts and their contacts, cutting off the disease's ability to spread. This approach, called ring vaccination, let a finite effort corner the virus outbreak by outbreak. The last natural case of the more severe form of smallpox was found in a young Bangladeshi woman named Rahima Banu in 1975, and the last natural case of any smallpox was a

Somali hospital worker named Ali Maow Maalin in 1977. In 1980, the World Health Organization declared smallpox eradicated. A disease that had killed hundreds of millions of people over centuries no longer exists in nature, surviving only in a couple of secure laboratory freezers. It stands as proof that a disease can, in principle, be eliminated from the world entirely through vaccination.

The campaign to eradicate polio has followed, and its story shows both how far this can go and how hard the last stretch is. When the Global Polio Eradication Initiative launched in 1988, polio was paralyzing an estimated 350,000 children a year across more than a hundred countries. Through decades of mass vaccination, largely with the oral polio vaccine, cases have fallen by more than ninety-nine percent, and the disease has been eliminated from almost the entire world. Two of the three types of wild poliovirus have been declared globally eradicated. But the last cases have proven stubborn, clinging to a small number of areas where conflict, difficult terrain, and mistrust have made vaccination hard, and wild polio has continued to circulate in parts of Afghanistan and Pakistan. The polio effort also illustrates the earlier point about live vaccines: the oral polio vaccine, while highly effective and easy to give, can in rare cases revert and cause vaccine-derived polio, which has complicated the endgame and driven the development of improved vaccines. Finishing the job of polio eradication has turned out to be far harder than getting most of the way there, which is a common pattern in global disease control.

Beyond these targeted eradication efforts, the broader work of vaccinating the world's children has quietly saved more lives than almost any other public health measure. In 1974, the World Health Organization launched what it called the Expanded Programme on Immunization, an effort to bring a core set of childhood vaccines to children everywhere, including in the poorest countries, against diseases like diphtheria, tetanus, whooping cough, measles, polio, and tuberculosis. At the time, only a small fraction of the world's children were vaccinated.

Today, most are, though coverage remains uneven and incomplete in many places. On the program's fiftieth anniversary, researchers estimated that childhood vaccination over that half-century had saved on the order of 154 million lives, the majority of them infants, with measles vaccination alone accounting for a large share of those saved. Numbers on this scale are hard to truly feel, but they represent an enormous reduction in childhood death, achieved largely out of public view.

Getting vaccines to everyone who needs them is a logistical challenge as much as a scientific one, and one obstacle in particular shapes the whole effort: the cold chain. Most vaccines must be kept cold, within a specific temperature range, from the moment they're manufactured until the moment they're given, or they lose their potency. Maintaining this unbroken chain of refrigeration is difficult enough in wealthy countries and far harder in hot climates, in remote areas without reliable electricity, and in places lacking good roads. Delivering vaccines to a remote village can require refrigerated trucks, solar-powered refrigerators, and insulated carriers packed with ice, all coordinated so the vaccine never gets too warm. The temperature sensitivity of live and mRNA vaccines makes this especially demanding. A great deal of the practical work of global immunization is really the work of keeping vaccines cold across long and difficult journeys, and improvements in heat-stable vaccines and better cold-chain technology have been quietly important to reaching more children.

Funding and cooperation have been essential to all of this, since the countries with the greatest need often have the fewest resources. Organizations created to bridge that gap, including a global vaccine alliance founded in 2000 that pools funding from governments and donors to buy vaccines for low-income countries, have helped immunize hundreds of millions of children who might otherwise have gone without. This kind of international cooperation, imperfect and often strained, is nonetheless a large part of why vaccination has reached the scale it has.

The work is far from finished, and the challenges ahead are real. Reaching the last unvaccinated children, often in conflict zones, remote areas, or marginalized communities, is the hardest part and the most expensive. Maintaining high vaccination rates in the face of complacency and misinformation is an ongoing struggle, and declining coverage has allowed diseases like measles to resurge even in wealthy countries that had nearly eliminated them. New vaccines continue to arrive, including the first vaccines against malaria, a parasitic disease that has resisted vaccination for a century and still kills hundreds of thousands of children a year, and these bring both promise and the challenge of deploying them widely. The history of global vaccination shows what's achievable when the world commits to it, and it also shows that the gains are never permanent, requiring continued effort, funding, and public trust to sustain.

Discussion Question

Herd immunity means an individual can sometimes skip a vaccine and still be protected by everyone else's immunity, which creates a tension between what benefits one person and what protects the whole community. Given that this same tension appears in many areas of life where individual and collective interests diverge, how should a society balance respect for personal choice against the reality that the vulnerable are protected only when enough people participate?

Chapter 9: Modern Immunology and Applications

For most of this book, immunology has been a story about understanding: how the immune system works, why it sometimes fails, and how vaccines borrow its logic.

This final chapter is about something different. It's about what happens when we stop merely understanding the immune system and start deliberately engineering it, turning it into a treatment for diseases it wasn't beating on its own.

Everything here builds on what you already know. Cancer immunotherapy uses the T cells and antibodies from earlier chapters. Transplant medicine is a running battle with the self versus non-self recognition you met on page one. The microbiome connects to tolerance and the training of your immune system. And personalized medicine ties it all together by tailoring these approaches to the individual. This is immunology at the frontier, where the science becomes medicine that saves lives today.

Immunotherapy for Cancer

In the chapter on immune dysfunction, you learned that cancer succeeds largely by escaping the immune system, and that many tumors survive by pulling the immune system's own brakes. Cancer immunotherapy is the effort to reverse that, to release the brakes, sharpen the attack, and turn a patient's own immune system into a treatment for their cancer. After more than a century of false starts, it has become one of the most successful new approaches in all of cancer medicine, and it's changed the outlook for patients who once had almost no options.

The idea is old, older than most people realize. In the 1890s, a New York surgeon named William Coley noticed that some cancer patients who developed serious bacterial infections saw their tumors shrink, apparently because the infection roused their immune system into attacking the cancer along with the microbes. He began deliberately injecting patients with bacterial preparations, later called Coley's toxins, and reported some remarkable results. But his approach was inconsistent and poorly understood, and as radiation and chemotherapy developed, immunotherapy was largely set aside for decades. Coley had glimpsed something real, but the science wasn't ready to explain or control it.

The modern breakthrough came from understanding the checkpoints, the molecular off switches that normally keep the immune system from overreacting. Recall from the chapter on dysfunction that T cells carry these checkpoints as a safety feature, and that cancers exploit them by displaying signals that engage the checkpoints and tell attacking T cells to stand down. Two scientists, working separately, worked out how to counter this. James Allison in the United States studied a checkpoint molecule called CTLA-4 and showed that blocking it could unleash T cells against tumors in mice. Tasuku Honjo in Japan discovered a different checkpoint called

PD-1 and its role in restraining T cells. Their work led to a class of drugs called checkpoint inhibitors, antibodies that block these off switches so the immune system can attack the cancer freely. Allison and Honjo shared the Nobel Prize in 2018 for opening this door.

Checkpoint inhibitors have produced results that genuinely surprised oncologists. The first one, ipilimumab, which blocks CTLA-4, was approved in 2011 for advanced melanoma, a skin cancer that was nearly a death sentence once it spread. Drugs that block PD-1, including pembrolizumab and nivolumab, followed and proved even more effective across a range of cancers. What made these drugs so striking wasn't just that they worked, but how they worked: in some patients with advanced, previously untreatable cancer, the tumors melted away and stayed gone for years, something chemotherapy almost never achieved. A widely noted example was former US President Jimmy Carter, who in 2015 was diagnosed with melanoma that had spread to his brain and liver, a situation that would once have been hopeless. After treatment that included pembrolizumab, his tumors disappeared, and his case became one of the most public demonstrations of what checkpoint inhibitors could do. Not everyone responds, and much current research aims at understanding why some patients benefit dramatically while others don't, but for those who do respond, the results can be lasting in a way older treatments rarely were.

Releasing the immune system's brakes has a predictable downside, and it illustrates a theme running through this whole book. If checkpoints normally prevent the immune system from attacking the body, then blocking them can let the immune system attack healthy tissue, essentially causing autoimmune side effects. Patients on checkpoint inhibitors can develop inflammation in the skin, gut, lungs, liver, and various glands, which sometimes requires treatment to suppress. It's the same trade-off you saw with autoimmunity and immunosuppression, viewed from the other side: turning the immune system up against cancer risks it turning against the patient. Managing that balance is part of the art of using these drugs.

A completely different immunotherapy takes the approach of engineering a patient's T cells directly, and it's one of the most remarkable things modern medicine can do. It's called CAR T-cell therapy, short for chimeric antigen receptor T-cell therapy, and the concept is extraordinary: doctors remove a patient's own T cells, genetically modify them in the laboratory to recognize the patient's cancer, grow them in large numbers, and infuse them back into the patient as a living drug. The modification adds a synthetic receptor, the chimeric antigen receptor, that's designed to lock onto a specific protein on the surface of the cancer cells. Once reinfused, these engineered T cells hunt down and kill cancer cells carrying that protein, and because they're living cells, they can multiply inside the patient and persist, continuing to patrol for the cancer.

The results in certain blood cancers have been stunning. CAR T-cell therapy was first developed against a protein called CD19 found on certain leukemia and lymphoma cells. One of the earliest and most famous patients was a six-year-old girl named Emily Whitehead, who in 2012 was dying of acute lymphoblastic leukemia that had resisted every standard treatment. She received an experimental CAR T-cell therapy, became gravely ill from the treatment, and then recovered completely, going into a remission that has now lasted well over a decade. Her case helped prove the approach, and in 2017 the first CAR T-cell therapy was approved for use. For patients with certain leukemias and lymphomas that had failed all other treatments, CAR T-cell therapy has produced remissions that would have been unimaginable a generation ago.

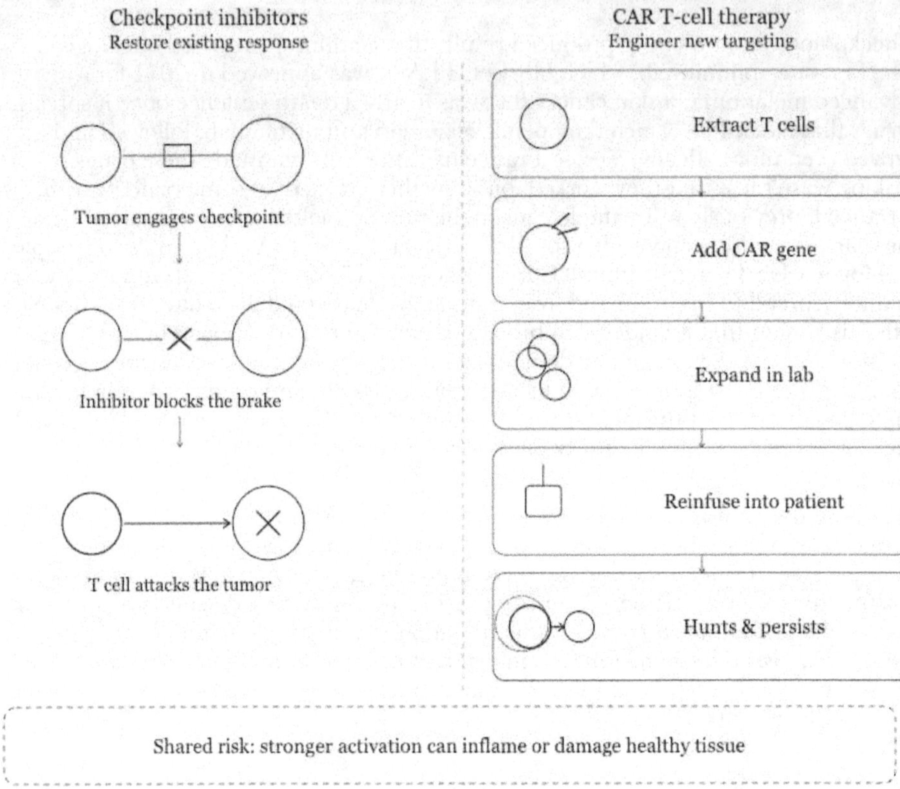

CAR T-cell therapy has real limitations, and being honest about them matters. The treatment can cause a severe reaction called cytokine release syndrome, in which the sudden, massive activation of the engineered T cells triggers a flood of inflammatory cytokines, the same kind of cytokine storm you met in the context of sepsis, which can cause dangerously high fevers, plummeting blood pressure, and organ stress. It's manageable with careful monitoring and specific treatments, including drugs that block the main cytokine involved, but it makes CAR T-cell therapy something that has to be given in specialized centers. The approach has also worked far better against blood cancers than against solid tumors like breast or lung cancer, which are harder for the engineered cells to penetrate and which surround themselves with defenses that shut the T cells down. And because each treatment is manufactured individually from a single patient's own cells, it's extremely expensive and complex to produce, which limits how widely it can be used. Researchers are working hard on all of these problems, trying to make CAR T-cell therapy work against solid tumors and to produce it more cheaply and quickly.

Cancer vaccines are another arm of immunotherapy, and here a distinction matters. Some vaccines prevent cancer by preventing the infections that cause it: the hepatitis B and HPV vaccines from the last chapter are, in effect, cancer vaccines, since they prevent viral infections that lead to liver and cervical cancers. Those are preventive vaccines given to healthy people. Therapeutic cancer vaccines are different: they're given to people who already have cancer, and they aim to train the immune system to attack the existing tumor. This has proven much harder, because a tumor that has already grown has already escaped the immune system once. One therapeutic cancer vaccine for advanced prostate cancer was approved back in 2010, showing the concept could work, though its benefits were modest.

The most exciting direction in cancer vaccines connects directly to the mRNA technology from the last chapter. Because a tumor's mutations produce abnormal proteins unique to that individual patient's cancer, called neoantigens, scientists can sequence a patient's tumor, identify the specific neoantigens, and rapidly design a personalized mRNA vaccine that trains the patient's immune system to recognize and attack their own particular cancer. This is a genuinely individualized medicine, a vaccine made for one person's tumor and no one else's. These personalized neoantigen vaccines are in clinical trials, often combined with checkpoint inhibitors, and early results, particularly in melanoma, have been encouraging enough to generate real optimism. It's one of the clearest examples of several threads in this book converging: understanding tumor mutations, the mRNA platform, and the immune system's ability to be taught, all in a single treatment.

Monoclonal antibodies, which you met in detail earlier, remain one of the workhorses of cancer immunotherapy, and they've grown more sophisticated. Beyond the checkpoint inhibitors, which are themselves monoclonal antibodies, and the older antibodies like trastuzumab and rituximab that target proteins on cancer cells, newer designs have added clever twists. Antibody-drug conjugates attach a potent toxin to a monoclonal antibody, so the antibody delivers the poison directly to the cancer cells it targets while largely sparing healthy tissue, essentially using the antibody as a guided missile carrying a chemotherapy payload. Bispecific antibodies are engineered to grab two different targets at once, and some are designed to grab a cancer cell with one arm and a T cell with the other, physically dragging the killer cell into contact with the tumor and triggering it to attack. These engineered antibodies extend the antibody biology from earlier chapters into increasingly precise cancer weapons.

Here's an overview of the main immunotherapy approaches and how they work.

Approach	How It Works	Best Successes So Far
Checkpoint inhibitors	Block the immune system's off switches so T cells can attack	Melanoma, lung, and many other cancers
CAR T-cell therapy	Engineer the patient's own T cells to target the cancer	Certain leukemias and lymphomas

Therapeutic cancer vaccines	Train the immune system against an existing tumor	Prostate cancer; neoantigen vaccines in trials
Monoclonal antibodies	Target cancer cells; deliver toxins; or recruit T cells	Breast cancer, lymphoma, and others
Cytokine therapy	Flood the body with immune-activating signals	Older approach; melanoma, kidney cancer

Cancer immunotherapy hasn't replaced surgery, radiation, and chemotherapy, and it doesn't work for every patient or every cancer. But it's added a fundamentally different strategy to cancer treatment, one based on the immune system rather than on directly poisoning or cutting out the tumor, and for some patients it has turned a fatal diagnosis into a manageable or even cured disease. It represents the practical payoff of understanding, at a deep level, how the immune system recognizes and attacks abnormal cells, and how cancers evade that attack.

Organ Transplant and Immune Rejection

Organ transplantation is one of modern medicine's great achievements, and it's also a permanent struggle against the immune system. Everything that makes your immune system good at its job, the ability to recognize foreign tissue and destroy it, works directly against a transplanted organ. The self versus non-self recognition from the first chapter is exactly what a transplant surgeon has to overcome, because to your immune system, a donated kidney or heart looks like an enormous foreign invader that must be attacked.

The core problem comes down to the HLA molecules you met early on, the personal identification tags that vary from person to person. When a donor organ is transplanted, your immune system reads the donor's HLA tags, recognizes that they don't match your own, and mounts a response to reject the organ. This is why the very first successful organ transplant, a kidney transplanted in Boston in 1954 by a surgeon named Joseph Murray, was performed between identical twins: because identical twins share the same HLA tags, the recipient's immune system saw the donated kidney as self and didn't reject it. Murray later received a Nobel Prize for this work. For everyone who isn't an identical twin, though, rejection was the great barrier, and overcoming it required two things: matching donors and recipients as closely as possible, and suppressing the recipient's immune system.

Rejection comes in several forms, distinguished by how fast they happen and what drives them, and knowing them clarifies why transplant patients are managed the way they are. Hyperacute rejection happens within minutes to hours of the transplant, and it's caused by antibodies the recipient already had against the donor's tissue, which attack the organ almost immediately and can destroy it before it ever functions. This catastrophic reaction is now largely prevented by careful matching and by testing the recipient's blood against the donor's tissue before surgery. Acute rejection happens over days to weeks and is driven mainly by T cells recognizing the foreign organ; it's the most common form and the main reason patients take immunosuppressive drugs, which control it well in most cases. Chronic rejection develops slowly over months and years, a gradual immune-driven damage that scars the organ's blood vessels and tissue, slowly degrading its function. Chronic rejection remains the hardest to prevent and is a major reason transplanted organs don't last forever.

Type of Rejection	Timing	Main Cause	Status
Hyperacute	Minutes to hours	Pre-existing antibodies against the donor	Largely preventable with matching and testing
Acute	Days to weeks	T cells attacking the foreign organ	Controlled by immunosuppressive drugs
Chronic	Months to years	Ongoing low-level immune damage	Still difficult to prevent

The advance that made transplantation broadly possible was the development of effective immunosuppressive drugs, and one discovery in particular changed everything. In the 1970s, a scientist named Jean-François Borel, working at a pharmaceutical company, studied a compound called cyclosporine, derived from a fungus, which turned out to powerfully suppress T cell responses. When cyclosporine came into use in the early 1980s, transplant survival rates jumped dramatically, because doctors could finally control acute rejection reliably. A related drug called tacrolimus, also derived from a microbe, followed and works similarly. These drugs, usually combined with steroids and other agents, are what allow a transplanted organ to survive in a recipient whose immune system would otherwise destroy it. Transplant patients typically take some combination of these immunosuppressants for the rest of their lives.

This brings us to the central, unavoidable trade-off of transplantation, which echoes the one you saw with autoimmune disease. To keep the immune system from rejecting the organ, you have to suppress it, but a suppressed immune system can't defend the body properly. So transplant recipients on lifelong immunosuppression face a higher risk of infections, including opportunistic ones that a healthy immune system would easily handle, and a higher risk of certain cancers, since the immune surveillance that normally catches developing cancers is dampened. The patient trades the risk of rejection for the risks of a weakened immune system, and managing that balance, enough suppression to protect the organ but not so much as to leave the patient dangerously vulnerable, is the ongoing work of transplant medicine. The goal that researchers are still pursuing is tolerance: a way to make the recipient's immune system specifically accept the donated organ as self, without suppressing the rest of the immune system, which would eliminate the trade-off entirely. Achieving reliable transplant tolerance would be a major advance, and it remains an active area of research.

There's a special complication that arises in bone marrow and stem cell transplants, and it's the mirror image of ordinary rejection. When you transplant bone marrow, you're transplanting immune cells themselves, and those donor immune cells can recognize the recipient's body as foreign and attack it. This is called graft-versus-host disease, and it's a serious problem in stem cell transplants, essentially the transplant attacking the patient rather than the patient attacking the transplant. It's managed with careful matching and immunosuppression, and interestingly, a mild version can even be beneficial in treating blood cancers, because the donor immune cells attack any remaining cancer cells along with the rest, an effect called graft-versus-leukemia.

The persistent shortage of donor organs has driven research into a striking possibility: using animal organs to replace human ones, an approach called xenotransplantation. Pigs have organs of roughly the right size, but a pig organ transplanted into a human would be rejected almost instantly and violently, because it carries obviously foreign molecules. Recent advances in gene editing have made it possible to modify pigs so their organs are less likely to be rejected, removing the pig molecules that trigger the fastest rejection and adding human genes to help the organ survive. In 2022, surgeons transplanted a gene-edited pig heart into a living human patient for the first time, and further experimental transplants of gene-edited pig hearts and kidneys into patients have followed. These are early, experimental steps, and the patients have faced serious challenges, but they represent a genuine attempt to solve the organ shortage by engineering around the immune barrier. Whether xenotransplantation becomes a routine reality remains to be seen, but it shows how far the effort to overcome immune rejection has come.

The Microbiome and Immune Regulation

One of the biggest shifts in how immunologists think about the immune system has come from realizing that it doesn't operate alone. Your body is home to a vast community of microorganisms, trillions of bacteria along with other microbes, living mostly in your gut, and this community, called the microbiome, turns out to be deeply involved in how your immune system develops and behaves. The old picture of the immune system as a defender against all microbes has given way to a more accurate one, in which the immune system lives in constant partnership with the helpful microbes that inhabit us.

The numbers are worth appreciating. Your body carries something on the order of tens of trillions of bacterial cells, roughly comparable to the number of your own cells, most of them packed into your large intestine. These aren't invaders your immune system is fighting; they're long-term residents, many of which help you, breaking down fibers you can't digest, producing certain vitamins, and crowding out dangerous microbes. Your immune system, remarkably, tolerates this enormous population, maintaining a careful peace with the helpful residents while staying ready to attack genuine threats. Managing that peace, deciding what to tolerate and what to attack right at the gut wall where the density of microbes is greatest, is one of the immune system's largest ongoing jobs.

Here's the finding that changed everything: the microbiome doesn't just coexist with your immune system, it helps build it. The clearest evidence comes from germ-free animals, raised in sterile conditions so they have no microbiome at all. These animals turn out to have poorly developed immune systems, with smaller and fewer immune structures in the gut, reduced numbers of certain immune cells, and abnormal immune regulation. Their immune systems never matured properly, because the microbes that normally train them were absent. When these germ-free animals are later given a normal microbiome, their immune systems develop more normally. This showed, unmistakably, that your resident microbes are essential teachers of your immune system, especially in early life.

What do the microbes teach? A great deal of it comes down to balance and tolerance, connecting directly to concepts from earlier chapters.

Certain gut bacteria promote the development of regulatory T cells, the immune system's brakes that suppress excessive responses and help maintain tolerance. Bacteria that ferment dietary fiber produce short-chain fatty acids, small molecules that support these regulatory T cells and help keep the gut lining healthy and its immune responses appropriately calm. In effect, a healthy microbiome helps tune the immune system toward tolerance and restraint, teaching it not to overreact. This is why the composition of your microbiome influences your risk of immune-related diseases, and it links back to the hygiene hypothesis from the chapter on dysfunction: the reduced microbial exposure of modern life may alter the microbiome in ways that leave the immune system poorly trained and more prone to the misfires of allergy and autoimmunity.

Microbiome Function	Effect on the Immune System
Trains immune development early in life	Immune structures and cells mature properly
Promotes regulatory T cells	Supports tolerance and restraint
Produces short-chain fatty acids from fiber	Keeps the gut lining and its immune responses healthy
Crowds out harmful microbes	Prevents dangerous bacteria from taking hold

When the microbiome is disrupted, a state called dysbiosis, problems can follow, and researchers are finding connections to a growing list of conditions. An imbalanced microbiome has been linked to inflammatory bowel diseases like Crohn's and ulcerative colitis, in which the immune system attacks the gut, and associations have been found with allergies, asthma, autoimmune diseases, obesity, and even conditions affecting the brain and mood through what's called the gut-brain axis. It's important to be careful here, because much of this research shows associations rather than proven cause and effect, and the field is young and still sorting out which microbiome changes cause disease versus which are merely along for the ride. But the sheer breadth of the connections has made the microbiome one of the most active areas in all of biology.

The practical applications are beginning to arrive, and one has already proven dramatically effective. A dangerous gut infection called Clostridioides difficile, often triggered when antibiotics wipe out the normal microbiome and let this harmful bacterium take over, can be extremely hard to cure and often recurs. The most effective treatment turned out to be restoring a healthy microbiome directly, through a fecal microbiota transplant, in which stool from a healthy donor, full of normal gut bacteria, is transferred into the patient's gut to reestablish a balanced community. It sounds crude, but it works, curing the great majority of these stubborn infections by crowding out the harmful bacterium with a restored, healthy microbiome. This success has spurred research into whether adjusting the microbiome could treat other conditions, from inflammatory bowel disease to metabolic problems, though most of that work is still experimental.

There's a fascinating connection back to cancer immunotherapy that shows how interconnected all of this is. Researchers discovered that the composition of a cancer patient's gut microbiome can influence how well they respond to checkpoint inhibitors, the immunotherapy drugs discussed earlier. Patients with certain gut bacteria respond better to these treatments than patients with a different microbial makeup, apparently because the microbiome shapes how active the immune system is. This has led to trials testing whether modifying the microbiome could make cancer immunotherapy work better in more patients. That the bacteria in your gut might affect whether an immunotherapy drug can beat your cancer is a striking illustration of how thoroughly the microbiome is woven into your immune function.

Immunology in Personalized Medicine

The thread running through this final chapter, and in many ways through the whole future of the field, is that immune treatments increasingly work best when they're tailored to the individual. Your immune system is as personal as your fingerprint, shaped by your particular HLA type, your unique history of exposures, your microbiome, and, in the case of cancer, the specific mutations in your tumor. Personalized medicine is the effort to match treatment to that individuality, and immunology is at the center of it.

You've already seen several examples without them being labeled as such. CAR T-cell therapy is inherently personalized, manufactured from each patient's own cells for that patient alone. Neoantigen cancer vaccines are custom-designed for one person's specific tumor mutations. These aren't one-size-fits-all drugs; they're treatments built around the individual, and they represent where a growing part of immune-based medicine is heading.

The personal nature of HLA, the identification tags from the first chapter, matters for treatment in ways beyond transplantation. Because HLA type varies so much between people and influences how the immune system reacts, a person's HLA type can predict dangerous reactions to certain drugs. A specific HLA variant, for instance, predicts a severe hypersensitivity reaction to an important HIV medication, so patients are now tested for that variant before being given the drug, and those who carry it receive a different medication instead. Another HLA variant predicts a dangerous skin reaction to a common seizure medication in certain populations, and testing for it prevents those reactions. This is personalized medicine in everyday practice: using a person's immune genetics to choose safer drugs and avoid harm before it happens.

In cancer treatment, matching therapy to the individual tumor has become central, and immunology drives much of it. Not every patient benefits from checkpoint inhibitors, so researchers have looked for biomarkers, measurable features that predict who will respond. Tumors that display more of the target molecule the drugs block, or that carry many mutations and therefore produce many abnormal proteins for the immune system to recognize, tend to respond better to checkpoint inhibitors. Doctors can test a patient's tumor for these features to help predict whether immunotherapy is likely to work, sparing patients treatments unlikely to help them and directing those treatments to patients more likely to benefit. In a notable milestone, one checkpoint inhibitor was approved for use against any cancer carrying a particular genetic feature, regardless of where in the body the tumor started, the first time a cancer treatment was approved based on a molecular marker rather than the organ affected. Treatment defined by the tumor's biology rather than its location is a genuine shift in how cancer is approached.

Where all of this is heading is a form of medicine that reads each person's immune system and tailors care accordingly. Imagine sequencing a patient's tumor to design a vaccine against its specific mutations, checking their immune biomarkers to select the immunotherapy most likely to work, considering their microbiome's effect on the response, and accounting for their HLA type to avoid dangerous drug reactions, all combined into a treatment plan built for that one individual. Pieces of this are already real, and the rest is the direction the field is moving. The deeper our understanding of how the immune system works in each person, the more precisely we can direct it, whether to attack a cancer, accept a transplant, or leave the body's own tissues in peace.

That's a fitting place to arrive, because it returns to where this book began. The immune system's fundamental task, running quietly beneath everything, is to tell what belongs in your body from what doesn't, and to respond with exactly the right force to exactly the right threats. Everything in modern immunology, every therapy and application in this chapter, is ultimately an effort to work with that task, sometimes correcting it when it errs, sometimes borrowing its logic, sometimes directing it against a new target. The more clearly we understand the system that has been protecting you your whole life, usually without a single sign that it's working, the better we become at helping it, and at stepping in when it needs help of its own.

Discussion Question

Nearly every immune-based treatment in this chapter runs into the same wall: helping the immune system do one job well often means interfering with its ability to do another. Unleashing it against cancer risks autoimmunity, suppressing it to protect a transplant opens the door to infection and cancer, and even the microbiome's calming influence can blunt a treatment meant to activate the immune system. Given that these trade-offs keep reappearing no matter which direction we push, do you think the goal of medicine should be to override the immune system's built-in balance, or to work within it, and what might that choice mean for the kinds of cures we can realistically hope for?

Historical Timeline of Immunology

Immunology grew from one stubborn observation, that survivors of certain diseases never caught them again, into one of the most active fields in all of science, and the entries below trace that path in order from the ancient world to the present.

Around 430 BC, during a terrible plague in Athens, the Greek historian Thucydides wrote down something no one could yet explain. He noticed that people who recovered from the disease could safely nurse the sick, because the same person was never struck down a second time. It's the first written account of what we now call acquired immunity, set down more than two thousand years before anyone understood how it worked.

Long before germ theory, people in China, India, and the Ottoman world had learned a dangerous but useful trick against smallpox. By at least the fifteenth and sixteenth centuries, Chinese practitioners were grinding dried smallpox scabs into a powder and blowing it into a healthy person's nose, while others rubbed material from a mild case into a small cut in the skin. The aim was to provoke a milder bout of the disease that left the survivor immune. This practice, called variolation, spread along trade routes across Asia, the Middle East, and Africa over the following centuries. It carried real risk, since you were deliberately giving someone smallpox and a fraction of those inoculated died, but the death rate was far lower than that of natural smallpox, and for a disease that killed or disfigured so many, the gamble often made sense.

The idea reached the English-speaking world in the early eighteenth century through two separate channels. In England, Lady Mary Wortley Montagu, who had survived smallpox herself and watched it kill her brother, saw variolation practiced in Constantinople and had her own children inoculated, then championed the method back home around 1721. That same year in Boston, during a raging smallpox epidemic, the minister Cotton Mather promoted inoculation after learning of it from an enslaved African man named Onesimus, and a physician named Zabdiel Boylston carried it out despite fierce public opposition.

The turning point came in 1796, in the hands of an English country doctor named Edward Jenner. Jenner had noticed a piece of rural folk wisdom: milkmaids who caught cowpox, a mild ailment picked up from cattle, seemed protected from smallpox. Acting on the hunch, he took material from a cowpox sore on the hand of a milkmaid named Sarah Nelmes and scratched it into the arm of an eight-year-old boy, James Phipps. The boy developed a mild cowpox reaction and recovered. Then Jenner did the thing no modern ethics board would allow: he deliberately exposed the boy to smallpox. Phipps didn't get sick. Jenner had shown that infection with a harmless relative could protect against a deadly disease, and with far less danger than variolation, since cowpox almost never killed. He named the technique after the Latin word for cow, vacca, giving us the word vaccine. It was the first true vaccine in history, and it launched everything that followed.

Jenner published his findings in 1798, and within a few years vaccination had spread across Europe and beyond.

For nearly a century, vaccination meant smallpox and nothing else, until the French chemist Louis Pasteur turned a single lucky observation into a general method. In 1879, Pasteur found by accident that an old, forgotten culture of chicken cholera bacteria had lost its ability to cause disease, yet still protected chickens against fresh, deadly cultures. He'd stumbled onto attenuation, the deliberate weakening of a pathogen. Pasteur applied the principle systematically, staging a famous public demonstration of an anthrax vaccine in sheep in 1881, and in 1885 using an experimental rabies vaccine to save a nine-year-old boy named Joseph Meister, who had been mauled by a rabid dog. Pasteur kept Jenner's word "vaccine" for all of these treatments as a tribute, even though none involved cows, and from then on vaccination became a strategy that could be aimed at one disease after another.

In 1882, the Russian zoologist Élie Metchnikoff pushed a rose thorn into a transparent starfish larva and watched mobile cells swarm to surround the intruder. He realized that certain cells actively engulf and devour invaders, a process he named phagocytosis, and he built from it the first cellular theory of immunity.

Just a few years later, in 1890, Emil von Behring and Kitasato Shibasaburō made a discovery that pointed in a different direction. They found that the blood serum of animals immunized against diphtheria or tetanus contained something that could neutralize the toxins of those diseases, and that transferring this serum to another animal transferred the protection. This was the birth of antibodies and of serum therapy, the first treatment that could actually save people dying of diphtheria. When the Nobel Prize in Physiology or Medicine was awarded for the very first time in 1901, it went to von Behring for this work.

Through the 1890s, the German scientist Paul Ehrlich developed the opposing view to Metchnikoff's, emphasizing the soluble antibodies in the blood rather than the cells, and his vision of a "magic bullet" that could target a disease precisely anticipated much of modern medicine.

In 1891, a New York surgeon named William Coley began injecting cancer patients with bacterial preparations after noticing that infections sometimes made tumors shrink. His "Coley's toxins" were the first real attempt at cancer immunotherapy, a century before the idea would be understood.

Around 1894, the Belgian scientist Jules Bordet discovered a component of blood serum that worked alongside antibodies to destroy bacteria, later named complement. He earned a Nobel Prize for the work in 1919.

In 1901, the Austrian physician Karl Landsteiner solved a mystery that had made blood transfusions a deadly lottery. He discovered that human blood comes in distinct types, now known as the ABO groups, and that mixing incompatible types causes the blood to clump because the immune system attacks the foreign cells. His finding made safe transfusion possible and was an early demonstration of self versus non-self recognition. He received the Nobel Prize for it in 1930.

In 1906, the Austrian pediatrician Clemens von Pirquet coined the word "allergy" to describe the immune system's altered, sometimes harmful reactivity.

In 1908, the Nobel Prize was shared by Metchnikoff and Ehrlich, honoring the two rival theories of immunity at once, the cellular and the humoral, in a fitting acknowledgment that the field needed both.

In 1922, the Scottish scientist Alexander Fleming discovered lysozyme, an enzyme in tears, saliva, and mucus that breaks down bacterial cell walls, when he added a drop of his own nasal mucus to a bacterial culture during a cold. It was one of the body's own antimicrobial weapons, found six years before Fleming stumbled onto penicillin.

During the 1920s, the French veterinarian Gaston Ramon worked out how to inactivate bacterial toxins so they could no longer harm the body but could still train it to defend itself, creating the first toxoid vaccines against diphtheria and tetanus. Ramon also pioneered the use of adjuvants, substances added to boost the immune response, an idea that remains central to vaccine design.

Landsteiner struck again in 1940, identifying the Rhesus (Rh) blood factor with Alexander Wiener, which explained a dangerous incompatibility between some mothers and their babies.

The problem of why the immune system tolerates the body's own tissues came into focus through the work of the British biologist Peter Medawar. During the Second World War, treating badly burned pilots, Medawar studied why skin grafts from donors were rejected, and he recognized rejection as an immune response. Working with Rupert Billingham and Leslie Brent, he then showed in 1953 that this rejection wasn't inevitable: mice exposed to foreign cells before birth grew up accepting grafts from that donor, a phenomenon they called acquired immunological tolerance. The experiments proved that the immune system learns during development which tissues to treat as self, and they laid the scientific groundwork for organ transplantation.

The theoretical foundation had been anticipated in 1949, when the Australian scientist Frank Macfarlane Burnet, with Frank Fenner, proposed that the body learns to distinguish self from non-self during development and predicted that tolerance could be induced, an idea Medawar's experiments soon confirmed.

In 1957, Burnet put forward clonal selection theory, one of the ideas the whole field now rests on. He proposed that the body generates an enormous diversity of lymphocytes before ever meeting an invader, each carrying a unique receptor, and that an antigen simply selects the cells that already match it and prompts them to multiply. It explained specificity, memory, and tolerance in a single framework.

Also in 1957, Alick Isaacs and Jean Lindenmann discovered interferon, the signaling protein that infected cells release to warn their neighbors and to rouse defenses against viruses.

Starting in 1958, the French immunologist Jean Dausset identified the human versions of the tissue-identity molecules, now known as the HLA system, by studying the reactions of patients who had received many blood transfusions. His work revealed the personal molecular tags that make each person's tissue distinct and that determine whether a transplant will be accepted or rejected.

Burnet and Medawar shared the 1960 Nobel Prize for the discovery of acquired immunological tolerance.

In 1961, the Australian-British scientist Jacques Miller discovered what the thymus actually does. Until then the organ was considered useless in adults, but Miller showed that it's where a whole class of lymphocytes matures, the cells we now call T cells, named for the thymus. His finding revealed a previously unrecognized arm of the immune system.

By 1965, Max Cooper and Robert Good, studying chickens, had separated the lymphocytes into two distinct populations: the thymus-derived T cells and the bursa-derived B cells, establishing the fundamental division of adaptive immunity that organizes the field to this day.

In 1972, Rodney Porter and Gerald Edelman shared the Nobel Prize for working out the structure of the antibody molecule. By cutting antibodies into fragments and analyzing their chains, they revealed the four-chain, Y-shaped architecture, with its target-grabbing arms and its signaling stem, that underlies everything antibodies do. Before their work, no one knew what an antibody looked like.

In 1973, Ralph Steinman discovered the dendritic cell, the branching sentinel that captures invaders and carries their fragments to the lymph nodes to launch the adaptive response. It would take decades for the field to grasp how central these cells are.

In 1974, Rolf Zinkernagel and Peter Doherty discovered MHC restriction, the rule that a T cell recognizes its target only when that target is displayed together with the body's own MHC molecules. The finding explained how T cells keep their attention on the body's own cells and only strike when one of them is displaying something foreign, and it reshaped the understanding of how T cells see the world.

In 1975, Georges Köhler and César Milstein, working in Cambridge, solved a problem that had frustrated scientists for years: how to produce a limitless supply of a single, identical antibody. They fused an antibody-producing B cell with a cancer cell to create a hybrid, called a hybridoma, that both made the exact antibody they wanted and divided forever. The resulting monoclonal antibodies could be manufactured in unlimited quantity, always the same, against any chosen target. The technique transformed biology and medicine, giving rise to a huge range of diagnostic tests and, eventually, to some of the best-selling drugs in the world. Köhler and Milstein shared a Nobel Prize for it in 1984.

Also around 1975, researchers working independently, including Rolf Kiessling and Ronald Herberman, identified natural killer cells, the innate lymphocytes that destroy compromised cells on sight without prior training.

The 1970s brought the single greatest triumph in the history of vaccination. A global campaign led by the World Health Organization, driven in large part by the

American epidemiologist Donald Henderson, chased smallpox around the world using surveillance and ring vaccination, cornering the virus outbreak by outbreak rather than trying to vaccinate everyone. The last natural case of the severe form was found in a young Bangladeshi woman, Rahima Banu, in 1975, and the last natural case of any smallpox appeared in a Somali hospital worker named Ali Maow Maalin in 1977. In 1980, the World Health Organization declared smallpox eradicated, the first and still the only human disease ever deliberately wiped from the Earth.

The 1980 Nobel Prize went to Baruj Benacerraf, Jean Dausset, and George Snell for uncovering the genetics of the MHC molecules that govern immune recognition.

In June 1981, physicians in the United States reported a cluster of strange, severe infections in previously healthy young men, the first recognized cases of what would become the AIDS epidemic. Two years later, in 1983, Françoise Barré-Sinoussi and Luc Montagnier at the Pasteur Institute isolated the virus responsible, HIV, which attacks and destroys the very helper T cells that coordinate the immune response. The discovery explained why the disease was so devastating and set off a decades-long scientific effort. Barré-Sinoussi and Montagnier received the Nobel Prize for identifying the virus in 2008.

In 1983, the immunosuppressive drug cyclosporine, derived from a fungus and shown by Jean-François Borel to powerfully suppress T cells, came into medical use and transformed organ transplantation. By reliably controlling rejection, it turned transplants from risky, often-fatal procedures into routine, life-saving operations.

In 1986, the first genetically engineered vaccine was approved: a hepatitis B vaccine made by programming yeast cells to manufacture a single viral protein, proving that a vaccine could be built from an isolated piece of a pathogen rather than the whole microbe.

In 1987, Susumu Tonegawa won the Nobel Prize for solving one of biology's deepest puzzles: how the body can make billions of different antibodies when it has only around twenty thousand genes. He'd shown that B cells don't store complete antibody genes but instead shuffle and recombine a modest set of gene fragments, generating almost unlimited diversity from a limited starting kit.

The year 1989 produced two ideas that would shape immunology for decades. Charles Janeway proposed that the adaptive immune system can't act on its own, arguing that the innate system must first recognize invaders through a set of built-in sensors and then grant the adaptive response permission to proceed. He predicted the existence of pattern recognition receptors years before they were found, and he correctly identified adjuvants as the hidden ingredient that makes vaccines work by triggering these innate sensors. That same year, the epidemiologist David Strachan proposed the hygiene hypothesis after noticing that children with more siblings had fewer allergies, suggesting that reduced exposure to microbes early in life leaves the immune system prone to misfiring.

In 1994, the immunologist Polly Matzinger proposed the danger model, arguing that the immune system responds not simply to anything foreign but to signals of damage and danger, a reframing that helped explain why the body tolerates harmless foreign things like food and gut bacteria.

In 1996, the French scientist Jules Hoffmann discovered that a gene called Toll, already known for its role in fruit fly development, was essential for the fly's defense against fungal infection, revealing an ancient recognition system that mammals turned out to share.

Two years later, in 1998, Bruce Beutler identified the mammalian sensor for bacterial LPS, one of the Toll-like receptors, confirming Janeway's prediction and pinning down how innate cells recognize a whole class of bacteria. The same year saw a darker episode: the British doctor Andrew Wakefield published a small, fraudulent study falsely linking the MMR vaccine to autism, a claim later thoroughly discredited but not before it damaged vaccination rates and public trust for years.

In 2002, Robert Schreiber and colleagues described cancer immunoediting, the three-phase process by which the immune system first eliminates cancer cells, then holds them in check, and finally, by killing the ones it can recognize, inadvertently selects for the variants able to escape.

In 2004, researchers discovered neutrophil extracellular traps, the sticky webs of DNA that dying neutrophils fling out to snare bacteria.

In 2005, Katalin Karikó and Drew Weissman published the discovery that would later make mRNA vaccines possible. They found that subtly modifying one of the building blocks of mRNA kept the body from treating it as a dangerous foreign signal, dramatically reducing inflammation while increasing how much protein the cells produced. The work drew little attention at the time and struggled for funding, but it proved essential fifteen years later.

In 2006, the first HPV vaccine was approved, protecting against the human papillomavirus that causes cervical and other cancers, and making it, in effect, a vaccine against cancer.

The year 2010 brought the approval of the first therapeutic cancer vaccine, sipuleucel-T for advanced prostate cancer, showing that the immune system could be trained to attack an existing tumor. The same year, the medical journal that had published Wakefield's vaccine-autism paper formally retracted it, and Wakefield lost his license to practice medicine.

In 2011, cancer immunotherapy came of age with the approval of ipilimumab, the first checkpoint inhibitor, a drug that releases one of the immune system's own brakes so that T cells can attack tumors. It produced lasting remissions in some patients with advanced melanoma, a cancer that had been nearly untreatable once it spread. That same year, the Nobel Prize honored the innate immune system, going to Bruce Beutler and Jules Hoffmann for their work on pattern recognition and to Ralph Steinman for discovering dendritic cells. In a poignant twist, Steinman had died of cancer just three days before the announcement, unaware he'd won, and the Nobel committee made the unusual decision to let the award stand.

In 2012, a six-year-old girl named Emily Whitehead, dying of leukemia that had resisted every standard treatment, became the first child to receive an experimental CAR T-cell therapy, in which her own T cells were genetically engineered to hunt her cancer. She fell gravely ill from the treatment and then recovered completely, entering a remission that has now lasted more than a decade, and her case helped prove that engineered immune cells could cure cancers nothing else could touch.

In 2015, former US President Jimmy Carter announced that melanoma which had spread to his brain and liver had disappeared after treatment that included the checkpoint inhibitor pembrolizumab, one of the most public demonstrations of what immunotherapy could achieve. The same year, a landmark study on peanut allergy showed that feeding peanuts to infants, rather than avoiding them, sharply reduced their chance of developing an allergy, overturning years of medical advice.

In 2017, the first CAR T-cell therapy was approved for use, and a checkpoint inhibitor became the first cancer drug ever approved based on a tumor's genetic features rather than the organ it started in.

In 2018, James Allison and Tasuku Honjo shared the Nobel Prize for discovering the immune checkpoints and showing how blocking them could unleash the immune system against cancer, the science behind checkpoint inhibitors.

In January 2020, Chinese scientists published the genetic sequence of a new coronavirus, SARS-CoV-2, as it began spreading around the world. Within days, researchers had designed mRNA vaccines against it, using the sequence alone and the modified-mRNA technology developed years earlier. After large clinical trials, the first mRNA vaccines were authorized in December 2020, roughly eleven months after the virus was identified, an unprecedented pace that made a decades-old technology suddenly famous and helped bring the pandemic under control.

In 2021, the World Health Organization recommended the first malaria vaccine for widespread use in children, a milestone against a parasitic disease that had resisted vaccination for a century and still kills hundreds of thousands of children each year.

In 2022, surgeons transplanted a genetically edited pig heart into a living human patient for the first time, an experimental step toward using animal organs to relieve the chronic shortage of human ones by engineering around immune rejection.

In 2023, Katalin Karikó and Drew Weissman received the Nobel Prize for the mRNA modifications that made the COVID-19 vaccines possible, an award that recognized research which had been overlooked and underfunded for much of its history before proving essential in a global emergency.

In 2024, the World Health Organization marked fifty years of its childhood immunization program, and researchers estimated that vaccination over that half-century had saved roughly 154 million lives, most of them infants, with measles vaccination alone accounting for a large share of those saved. It was a reminder that the quiet, ongoing work of vaccinating the world's children has prevented more death than almost any other measure in the history of medicine.

End of Chapter Discussion Questions

Chapter 1 - How should the body set the balance between responding too weakly and too strongly, and what makes it drift over a lifetime?

The body doesn't decide consciously. It settles the balance through layered checks that each demand confirmation before a response escalates. A single detection rarely triggers full action; cells require corroborating danger signals, and regulatory cells actively push back against overreaction. This design keeps the resting state calm and reserves force for genuine threats. The balance drifts for concrete reasons. Aging wears down the thymus and dulls regulation, so older people face both weaker defense and more autoimmunity at once. Chronic stress, poor sleep, and excess body fat nudge the system toward constant low-grade inflammation. Losing regulatory cells or breaking tolerance tips it toward attacking the self. A serious infection can retune it for months afterward. The honest answer is that there's no single correct setting. The body aims for a moving target and pays a price whenever it lands too far to either side.

Chapter 2 - Why keep two cell types that kill the same way but on different timelines instead of just one?

Because the two cells solve different halves of a timing problem, and no single cell could cover both. Speed and precision pull against each other. Natural killer cells act in the first hours, before your body knows what it's facing, and they strike any cell that has dropped its identity tags. That immediacy is priceless early in a viral infection, but it comes at a cost: NK cells work off a crude rule and can't tell one virus from another. Cytotoxic T cells fix that. They take days to activate because they have to find the rare matching cell, confirm the exact target, and multiply, but the result is a force aimed at one specific invader, plus lasting memory for next time. A body with only NK cells would respond fast but bluntly and forget everything. A body with only cytotoxic T cells would be helpless during the crucial first days while the slow buildup ran. Keeping both means you're covered across the whole timeline. Viruses also exploit this: hiding from one killer often exposes you to the other, so running both closes an escape route.

Chapter 3 - Why did evolution favor an aggressive, self-harming inflammatory response?

Because for most of human history, dying of infection was a far more immediate threat than any slow disease inflammation might cause decades later. Evolution optimizes for surviving to reproduce, not for a comfortable old age. In a world of frequent wounds, contaminated food, and no antibiotics, a fast, brutal response that stopped an infection in hours was worth the tissue damage, even when it sometimes went too far. The chronic diseases inflammation now drives mostly strike later in life, after the reproductive years, so there was little pressure to soften it. Our defenses are tuned for a dangerous past, not our safer, longer-lived present.

Chapter 4 - Why prepare a vast repertoire of lymphocytes for threats you may never meet, and what does that reveal about how the two systems bet on the future?

It makes sense because the future is unknowable, and the cost of being unprepared for the one threat that arrives is far higher than the cost of maintaining cells that sit idle. You can't predict which virus you'll meet next year, or whether some entirely new microbe will emerge, so the adaptive system refuses to bet on any particular enemy. Instead it prepares for nearly all of them at once, generating so much diversity that a rough match already exists for almost anything. Most of those cells will die unused. That's the price of insurance, and it's a bargain next to facing a lethal invader with no matching defender. This reveals a deep difference in strategy. The innate system bets on the past: it targets the molecular features microbes have shared for hundreds of millions of years, patterns so essential that invaders can't abandon them. It's cheap and reliable precisely because it wagers only on what has always been true. The adaptive system bets on the unknown future, spending resources up front so it can recognize threats that don't exist yet. One system says the enemy will look like it always has, and it's usually right. The other says the enemy might look like anything, and prepares accordingly. Together they hedge across both certainty and surprise, which is exactly what you'd want when survival depends on being ready for a world you can't see coming.

Chapter 5 - Why is a randomly built antibody system, which can bind self, a better bet than one that builds only the "right" antibodies?

Because you can't build only the right antibodies when you don't know what "right" will be. To pre-specify the correct ones, the body would need to know every invader in advance, including microbes that haven't evolved yet and synthetic molecules that never existed in nature. That's impossible. Randomness sidesteps the problem entirely: by shuffling gene fragments blindly, the body generates a match for almost anything without needing to anticipate it. The risk of self-binding is real, but the body handles it with screening that deletes self-reactive cells during development and brakes that suppress the ones that slip through. That's a solvable safety problem. Being blind to a novel threat is not.

Chapter 6 - If feeling sick often means your defenses are working, how should that change how we treat symptoms, and where's the line?

It should make us more thoughtful about reflexively suppressing every symptom, because some of what feels like illness is the cure at work. A mild fever slows many microbes while speeding your immune cells, so knocking it down with medication may trade a little discomfort for a slightly slower recovery. The same logic applies to the fatigue that pushes you to rest and the inflammation that walls off an infection. None of this means suffering is virtuous or that symptoms should always be endured. The line falls where the response stops helping and starts threatening. A high fever in a young child that risks seizures, a temperature climbing to dangerous levels, breathing compromised by swelling, or inflammation severe enough to damage organs: these are responses that have overshot their usefulness and need controlling. For a healthy adult with an ordinary fever, the calculation tilts toward letting a mild response run while treating yourself for comfort. The practical rule is to suppress symptoms when they endanger you or when the misery clearly outweighs the modest benefit, but don't assume that feeling bad always means something is going wrong. Sometimes it means something is going right.

Chapter 7 - Would an immune system tuned more aggressive, or one tuned less, be better, and what does that say about where evolution settled?

Neither setting is clearly better, which is exactly why the question is hard and why evolution landed in the uncomfortable middle. Turn the system up and you'd catch more early cancers, but you'd pay in autoimmunity, attacking joints, nerves, and organs in far more people. Turn it down and you'd spare people autoimmune disease, but more cancers would slip through and more infections would take hold. Each direction trades one kind of suffering for another.

Evolution's settling point reflects a specific bargain: it favored whatever balance let the most people survive to reproduce, and for most of history that meant prioritizing defense against infection over both cancer and autoimmunity, since infections killed the young while cancer mostly killed the old. The result isn't optimized for any individual or for modern lifespans. It's a compromise tuned for ancestral conditions, which is part of why both autoimmune disease and cancer persist. Perfect discrimination may be impossible, so the body was left holding a setting that fails in both directions rather than one that fails in neither.

Chapter 8 - How should a society balance personal choice against the fact that the vulnerable are protected only when enough people vaccinate?

This is one of the clearest cases where individual freedom and collective welfare genuinely collide, and there's no frictionless answer. The core tension is real. Herd immunity is a shared resource, built by many people accepting a small individual cost, and it protects exactly those who can't protect themselves: the newborn, the cancer patient, the person whose immune system can't respond. When enough people opt out, that shared protection collapses, and the harm falls hardest on the vulnerable who had no say in the matter. So a purely individual framing misses something important, because your vaccination decision affects other people's bodies, not just your own.

That said, coercion carries its own costs, and heavy-handed mandates can erode the public trust that vaccination depends on, sometimes backfiring. Most societies settle on a layered approach that mirrors how we handle other collective-risk situations. They make vaccination easy, free, and the default. They require it for shared settings like schools, where one person's choice most directly endangers others, while allowing genuine medical exemptions. They rely mainly on education and trust rather than force. The honest position is that neither absolute freedom nor absolute compulsion fits. The vulnerable have a real claim on the herd's protection, and the rest of us have a real claim to bodily autonomy, and a decent society keeps adjusting the balance rather than pretending either one wins outright.

Chapter 9 - Should medicine override the immune system's built-in balance or work within it, and what does that mean for the cures we can hope for?

The most durable successes will likely come from working within the balance rather than overriding it, though medicine will keep doing both. Overriding works, but the trade-offs never vanish: unleash the system and it turns on the body, suppress it and infection creeps in. Those aren't bugs to be brute-forced away; they're the cost of blunt tools. The more promising path is precision, teaching the system to accept one transplanted organ while defending against everything else, or to attack one tumor without harming healthy tissue. That's not overriding the balance but retuning it, target by target. Cures that respect the system's logic, rather than fighting it, are the ones most likely to last.

Conclusion

You started this book with a paper cut and a body that fixed it without asking. Now you know what was happening under the skin: the barriers that block most invaders, the fast innate cells that swarm a wound, the slow adaptive response that learns an enemy and remembers it for years. What looked like a single act of healing turns out to be a whole system, layered and coordinated, running quietly beneath everything you do.

A few ideas have come up again and again, and they're worth carrying with you. The first is the problem of self versus non-self. Almost everything your immune system does comes down to telling your own cells from foreign ones, and responding to real damage rather than harmless things like food and gut bacteria. Get that distinction right, and you stay healthy. Get it wrong in one direction and you get autoimmune disease. Get it wrong in the other and infections or cancer take hold.

The second idea is balance. Your immune system is powerful enough to kill your own cells, and that power cuts both ways. Nearly every disease in this book comes from the system doing too much or too little: overreacting in allergies, attacking the body in autoimmunity, smoldering in chronic inflammation, or failing in immunodeficiency. Even the treatments run into it. Unleash the immune system against cancer and you risk autoimmunity. Suppress it to protect a transplant and you open the door to infection. There's no setting that avoids every failure, only trade-offs the body has learned to manage.

The third idea is memory, and it's the one that changed the world. Because your adaptive system remembers the invaders it defeats, a first hard fight becomes an easy second one. That single fact is the foundation of vaccination, which has saved more lives than almost any other thing humans have ever done, and it's now being turned against cancer itself.

What should stay with you most is how much of this you now understand from the inside. When you get a vaccine, you know it's producing memory cells, not fighting an infection. When you run a fever, you know your body raised the temperature on purpose. When you hear about a new immunotherapy, an organ transplant, or a study on the microbiome, you have the framework to make sense of it. The headlines aren't jargon anymore. They're chapters of a story you've just read.

Immunology is moving fast, and much of what's exciting today didn't exist a decade ago. New answers will keep arriving, and some of what's in this book will be refined as researchers learn more. But the foundations don't change. Cells, antibodies, recognition, memory, and balance are the pieces everything else is built from, and you have them now.

Your immune system will keep protecting you every hour of every day, usually without a single sign that it's working. The difference is that now you know it's there, and you know a little of how it does what it does. You may never see the battles it wins, but every ordinary day is evidence that it has won them.

Key Terms and Definitions

- **Immune system** - The distributed network of cells, tissues, and organs that defends the body against threats, clears away damaged cells, and watches for cancer. It isn't a single organ but a coordinated system spread throughout the body.
- **Innate immunity** - The fast, general defense you're born with. It responds within minutes to hours, recognizes broad categories of threat, and keeps no memory.
- **Adaptive immunity** - The slower, highly specific defense that learns a particular invader, targets it precisely, and remembers it for the future. It's found only in vertebrates.
- **Pathogen** - Any organism that causes disease, including viruses, bacteria, fungi, and parasites.
- **Antigen** - Any molecule the immune system can recognize and respond to. Most antigens are proteins or fragments of proteins from an invader.
- **Immunogen** - An antigen that actually provokes an immune response, as opposed to one the system can merely recognize.
- **Antibody (immunoglobulin)** - A Y-shaped protein made by B cells that binds a specific target, marking it, blocking it, or flagging it for destruction.
- **Epitope** - The specific small patch on an antigen that an antibody or receptor actually binds to. A single large antigen carries many epitopes.
- **Paratope** - The part of an antibody, at the tip of its arm, that grips the epitope.
- **Self versus non-self** - The immune system's core task of distinguishing the body's own cells from foreign material.
- **Tolerance** - The trained ability of the immune system to leave the body's own tissues alone. Central tolerance removes self-reactive cells during development; peripheral tolerance restrains any that escape.
- **Phagocytosis** - The process by which a cell engulfs and digests an invader or debris.
- **Neutrophil** - The most abundant white blood cell and the fastest first responder, which engulfs bacteria and dies in large numbers during infection.
- **Macrophage** - A large, long-lived phagocyte that patrols tissues, cleans up debris, drives inflammation, and later helps repair damage. It takes on specialized forms in different organs.
- **Monocyte** - A circulating cell that travels in the blood and matures into a macrophage once it settles into a tissue.
- **Dendritic cell** - A branching sentinel cell that captures invaders and carries their fragments to the lymph nodes, launching the adaptive response.
- **Antigen presentation** - The display of antigen fragments on a cell's surface, held by MHC molecules, so that T cells can recognize them.
- **MHC (major histocompatibility complex)** - The surface molecules that display antigen fragments and act as identity tags on cells. In humans these are called HLA.
- **HLA (human leukocyte antigen)** - The human version of MHC. HLA genes are the most variable in the human genome, making each person's tissue nearly unique.

- **MHC class I** - Display molecules found on nearly all cells, showing samples of what a cell is making inside. Read by cytotoxic T cells.
- **MHC class II** - Display molecules found only on specialized antigen-presenting cells, showing invaders captured from outside. Read by helper T cells.
- **T lymphocyte (T cell)** - An adaptive immune cell that matures in the thymus and recognizes antigens displayed on MHC molecules.
- **Helper T cell (CD4)** - The coordinator of the immune response, which directs B cells, killer cells, and other defenders through chemical signals.
- **Cytotoxic T cell (CD8)** - The killer T cell, which destroys infected or cancerous cells displaying abnormal fragments on MHC class I.
- **Regulatory T cell (Treg)** - A T cell that suppresses and controls immune responses, acting as the brakes and maintaining tolerance.
- **B lymphocyte (B cell)** - An adaptive immune cell that matures in the bone marrow, recognizes invaders directly, and produces antibodies.
- **Plasma cell** - An activated B cell that has become an antibody factory, secreting large numbers of antibodies.
- **Natural killer cell (NK cell)** - An innate lymphocyte that destroys cells missing their MHC tags, acting immediately without prior training.
- **Memory cell** - A long-lived T or B cell that persists after an infection and enables a faster, stronger response to the same invader later.
- **Cytokine** - A signaling molecule released by immune cells to communicate, recruit other cells, and shape the response.
- **Interferon** - A cytokine released by virus-infected cells that warns neighboring cells and rouses antiviral defenses.
- **Complement system** - A group of more than thirty blood proteins that tag invaders, drive inflammation, and can puncture microbes directly.
- **Opsonization** - The coating of an invader with antibodies or complement so that phagocytes can catch and eat it more easily.
- **Membrane attack complex** - A ring of complement proteins that punches a hole in a microbe's membrane, causing it to burst.
- **Inflammation** - The coordinated response at a site of injury or infection, marked by redness, heat, swelling, and pain, that brings defenders to the area.
- **Histamine** - A chemical released by mast cells that widens and loosens blood vessels, driving redness and swelling. It's a main cause of allergy symptoms.
- **Mast cell** - A tissue-stationed immune cell that releases histamine when triggered, involved in inflammation and allergic reactions.
- **Eosinophil** - A white blood cell that attacks large parasites like worms by releasing toxic contents onto them, and that contributes to allergies.
- **Chemotaxis** - The movement of immune cells along a chemical trail toward the site of infection or injury.
- **Pattern recognition receptor (PRR)** - A sensor on innate immune cells that detects molecular signatures shared by classes of microbes.
- **PAMP (pathogen-associated molecular pattern)** - A molecular fingerprint common to many microbes but absent from the body, which PRRs recognize.
- **DAMP (damage-associated molecular pattern)** - A distress signal released by the body's own injured cells, which also triggers innate alarm receptors.
- **Toll-like receptor** - A well-known family of pattern recognition receptors that detects a range of microbial patterns.

- **Inflammasome** - A protein complex that, when triggered, releases powerful inflammatory signals and ramps up the response quickly.
- **Clonal selection** - The principle that the body makes a vast diversity of lymphocytes in advance, and an invader selects the ones that already match it.
- **Clonal expansion** - The rapid multiplication of a matching lymphocyte into many identical copies after activation.
- **Costimulation** - A second confirming signal required to fully activate a T cell, ensuring it acts only when there's genuine danger.
- **Anergy** - A shut-down, unresponsive state a lymphocyte enters when it recognizes an antigen without a danger signal, helping prevent autoimmunity.
- **Somatic hypermutation** - Rapid mutation of antibody genes in activated B cells, generating variants that may bind the invader more tightly.
- **Affinity maturation** - The process by which B cells producing better-fitting antibodies are selected over rounds, improving antibody quality during an infection.
- **Class switching** - A B cell's ability to change which class of antibody it produces while keeping the same target specificity.
- **Germinal center** - A structure in lymph nodes where activated B cells refine and switch their antibodies.
- **Affinity** - The strength of binding between a single antibody arm and a single epitope.
- **Avidity** - The total binding strength when an antibody grips multiple targets at once with multiple arms.
- **IgM** - The first antibody class produced in an infection. It joins into clusters of five and strongly activates complement.
- **IgG** - The most abundant antibody in the blood and the main all-purpose class. It crosses the placenta to protect newborns.
- **IgA** - The antibody of secretions such as saliva, tears, and mucus. It guards surfaces and is passed in breast milk.
- **IgE** - The antibody that fights large parasites and drives allergic reactions by triggering mast cells.
- **IgD** - The least understood antibody class, found on the surface of naive B cells.
- **Neutralization** - The blocking of an invader by antibodies so it can't attach to and enter the body's cells.
- **Immunological memory** - The lasting record of a defeated invader, held by memory cells and long-lived plasma cells, enabling rapid future defense.
- **Primary immune response** - The slower response on first exposure to an invader, taking days to build up.
- **Secondary immune response** - The faster, stronger response on later exposure, driven by memory cells.
- **Active immunity** - Long-lasting immunity the body builds itself by mounting a response and forming memory.
- **Passive immunity** - Immediate but temporary immunity gained by receiving ready-made antibodies from another source.
- **Vaccine** - A preparation that safely exposes the immune system to an antigen so it builds memory without causing the disease.
- **Attenuation** - The deliberate weakening of a pathogen so it can provoke immunity without causing illness.
- **Adjuvant** - A substance added to a vaccine to heighten the immune response to its antigen.

- **Conjugate vaccine** - A vaccine that links a bacterial sugar to a protein, recruiting T cell help so it works even in infants.
- **Toxoid** - An inactivated bacterial toxin used in vaccines to train the body to make neutralizing antitoxin antibodies.
- **mRNA vaccine** - A vaccine that delivers genetic instructions telling the body's own cells to make a target antigen.
- **Herd immunity** - The protection of a whole community, including the non-immune, when enough people are immune to stop a disease from spreading.
- **Basic reproduction number (R-nought)** - The average number of people one infected person would infect in a fully susceptible population; a measure of contagiousness.
- **Autoimmune disease** - A condition in which the immune system attacks the body's own tissues, mistaking self for a threat.
- **Molecular mimicry** - When a microbe carries a molecule resembling one of the body's own, causing antibodies against the microbe to attack the body.
- **Allergen** - A normally harmless substance, such as pollen or peanut protein, that triggers an allergic reaction.
- **Anaphylaxis** - A severe, body-wide allergic reaction that can cause collapsing blood pressure and blocked airways within minutes.
- **Immunodeficiency** - A state in which part of the immune system is missing or too weak, leaving the body open to infection.
- **SCID (severe combined immunodeficiency)** - A severe inherited condition in which both the T cell and B cell arms of adaptive immunity are absent or nonfunctional.
- **Immunosenescence** - The gradual weakening of the immune system with age.
- **Chronic inflammation** - Long-lasting, low-grade inflammation that never resolves and slowly damages tissue, linked to many serious diseases.
- **Immune surveillance** - The ongoing patrol by which the immune system finds and destroys cancer cells before they form tumors.
- **Immunoediting** - The process by which the immune system, in attacking a cancer, selects for the variants best able to escape it.
- **Checkpoint inhibitor** - A drug that blocks the immune system's off switches so T cells can attack a tumor freely.
- **CAR T-cell therapy** - A treatment in which a patient's own T cells are genetically engineered to recognize and kill their cancer.
- **Monoclonal antibody** - A uniform antibody produced from a single B cell clone, targeting one specific epitope, used widely as drugs and tests.
- **Neoantigen** - An abnormal protein produced by a tumor's mutations, unique to that patient's cancer, used as a target for personalized vaccines.
- **Immunosuppression** - The deliberate dampening of the immune system, used to control autoimmunity and prevent transplant rejection.
- **Graft-versus-host disease** - A complication of bone marrow transplants in which the donor's immune cells attack the recipient's body.
- **Xenotransplantation** - The transplantation of organs from another species, such as gene-edited pigs, into humans.
- **Microbiome** - The vast community of microbes living in and on the body, mostly in the gut, that helps train and regulate the immune system.
- **Dysbiosis** - An unhealthy imbalance in the microbiome, linked to a range of immune and other diseases.

References & Recommended Readings

Works Cited

Abbas, A. K., Lichtman, A. H., & Pillai, S. (2022). *Cellular and molecular immunology* (10th ed.). Elsevier.

Barré-Sinoussi, F., Chermann, J. C., Rey, F., Nugeyre, M. T., Chamaret, S., Gruest, J., … Montagnier, L. (1983). Isolation of a T-lymphotropic retrovirus from a patient at risk for acquired immune deficiency syndrome (AIDS). *Science, 220*(4599), 868–871.

Billingham, R. E., Brent, L., & Medawar, P. B. (1953). Actively acquired tolerance of foreign cells. *Nature, 172*(4379), 603–606.

Burnet, F. M. (1957). A modification of Jerne's theory of antibody production using the concept of clonal selection. *The Australian Journal of Science, 20*, 67–69.

Du Toit, G., Roberts, G., Sayre, P. H., Bahnson, H. T., Radulovic, S., Santos, A. F., … Lack, G. (2015). Randomized trial of peanut consumption in infants at risk for peanut allergy. *New England Journal of Medicine, 372*(9), 803–813.

Dunn, G. P., Bruce, A. T., Ikeda, H., Old, L. J., & Schreiber, R. D. (2002). Cancer immunoediting: From immunosurveillance to tumor escape. *Nature Immunology, 3*(11), 991–998.

Edelman, G. M. (1959). Dissociation of γ-globulin. *Journal of the American Chemical Society, 81*(12), 3155–3156.

Fenner, F., Henderson, D. A., Arita, I., Ježek, Z., & Ladnyi, I. D. (1988). *Smallpox and its eradication*. World Health Organization.

Isaacs, A., & Lindenmann, J. (1957). Virus interference. I. The interferon. *Proceedings of the Royal Society of London. Series B, 147*(927), 258–267.

Ishida, Y., Agata, Y., Shibahara, K., & Honjo, T. (1992). Induced expression of PD-1, a novel member of the immunoglobulin gene superfamily, upon programmed cell death. *The EMBO Journal, 11*(11), 3887–3895.

Jenner, E. (1798). *An inquiry into the causes and effects of the variolae vaccinae*. Sampson Low.

Karikó, K., Buckstein, M., Ni, H., & Weissman, D. (2005). Suppression of RNA recognition by Toll-like receptors: The impact of nucleoside modification and the evolutionary origin of RNA. *Immunity, 23*(2), 165–175.

Köhler, G., & Milstein, C. (1975). Continuous cultures of fused cells secreting antibody of predefined specificity. *Nature, 256*(5517), 495–497.

Leach, D. R., Krummel, M. F., & Allison, J. P. (1996). Enhancement of antitumor immunity by CTLA-4 blockade. *Science, 271*(5256), 1734–1736.

Lemaitre, B., Nicolas, E., Michaut, L., Reichhart, J. M., & Hoffmann, J. A. (1996). The dorsoventral regulatory gene cassette spätzle/Toll/cactus controls the potent antifungal response in Drosophila adults. *Cell, 86*(6), 973–983.

Matzinger, P. (1994). Tolerance, danger, and the extended family. *Annual Review of Immunology, 12*, 991–1045.

Miller, J. F. A. P. (1961). Immunological function of the thymus. *The Lancet, 278*(7205), 748–749.

Murphy, K., & Weaver, C. (2016). *Janeway's immunobiology* (9th ed.). Garland Science.

Poltorak, A., He, X., Smirnova, I., Liu, M. Y., Van Huffel, C., Du, X., … Beutler, B. (1998). Defective LPS signaling in C3H/HeJ and C57BL/10ScCr mice: Mutations in Tlr4 gene. *Science, 282*(5396), 2085–2088.

Porter, R. R. (1959). The hydrolysis of rabbit γ-globulin and antibodies with crystalline papain. *Biochemical Journal, 73*(1), 119–126.

Punt, J., Stranford, S. A., Jones, P. P., & Owen, J. A. (2018). *Kuby immunology* (8th ed.). W. H. Freeman.

Shattock, A. J., Johnson, H. C., Sim, S. Y., Carter, A., Lambach, P., Hutubessy, R. C. W., … Bhutta, Z. A. (2024). Contribution of vaccination to improved survival and health: Modelling 50 years of the Expanded Programme on Immunization. *The Lancet, 403*(10441), 2307–2316.

Steinman, R. M., & Cohn, Z. A. (1973). Identification of a novel cell type in peripheral lymphoid organs of mice. I. Morphology, quantitation, tissue distribution. *Journal of Experimental Medicine, 137*(5), 1142–1162.

Strachan, D. P. (1989). Hay fever, hygiene, and household size. *BMJ, 299*(6710), 1259–1260.

Tonegawa, S. (1983). Somatic generation of antibody diversity. *Nature, 302*(5909), 575–581.

von Behring, E., & Kitasato, S. (1890). Über das Zustandekommen der Diphtherie-Immunität und der Tetanus-Immunität bei Thieren. *Deutsche Medizinische Wochenschrift, 16*, 1113–1114.

Zinkernagel, R. M., & Doherty, P. C. (1974). Restriction of in vitro T cell-mediated cytotoxicity in lymphocytic choriomeningitis within a syngeneic or semiallogeneic system. *Nature, 248*(5450), 701–702.

Recommended Readings

Biss, E. (2014). *On immunity: An inoculation.* Graywolf Press.

A thoughtful, beautifully written look at why vaccination stirs such deep fears, blending science with history and culture. It's the best book for understanding the human side of the vaccine debate.

Davis, D. M. (2018). *The beautiful cure: The revolution in immunology and what it means for your health.* University of Chicago Press.

Tells the story of the recent discoveries that made immunotherapy possible.

Graeber, C. (2018). *The breakthrough: Immunotherapy and the race to cure cancer.* Twelve.

A gripping account of how scientists learned to turn the immune system against cancer, following the patients and researchers behind checkpoint inhibitors and CAR T cells.

Mukherjee, S. (2010). *The emperor of all maladies: A biography of cancer.* Scribner.

This Pulitzer Prize–winning history of cancer gives essential background for understanding why immunotherapy matters so much. It's long but consistently rewarding.

Sompayrac, L. M. (2019). *How the immune system works* (6th ed.). Wiley-Blackwell.

A short, famously clear primer that many medical students read alongside their textbooks. Turn to it if you want the core concepts explained again from a different angle.

Yong, E. (2016). *I contain multitudes: The microbes within us and a grander view of life.* Ecco.

A lively tour of the microbiome and how the microbes inside us shape our health, including our immune systems.